Graph Neural Networks for Neurological Disorders

Md. Mehedi Hassan • Anindya Nag
Shariful Islam • Herat Joshi
Editors

Graph Neural Networks for Neurological Disorders

Fundamentals, Applications and Benefits in Research and Diagnostics

Springer

Editors
Md. Mehedi Hassan
UniSA STEM
University of South Australia
Mawson Lakes, SA, Australia

Shariful Islam
Physical Activity and Nutrition
Deakin University
Burwood, VIC, Australia

Anindya Nag
Department of Computer Science
and Engineering
Northern University of Business
and Technology Khulna
Khulna, Bangladesh

Herat Joshi
Analytics and Decision Support
Great River Health Systems
West Burlington, IA, USA

ISBN 978-3-032-04314-6 ISBN 978-3-032-04315-3 (eBook)
https://doi.org/10.1007/978-3-032-04315-3

This Springer imprint is published by the registered company Springer Nature Switzerland AG
The registered company address is: Gewerbestrasse 11, 6330 Cham, Switzerland

Preface

Chapter 1 explores the fundamentals of GNNs, a powerful machine learning technique for modeling complex relationships in graph-structured data. We cover core architectures such as GCNs, GATs, and GraphSAGE, along with advanced models like GINs and R-GCNs. Additionally, we discuss key optimization techniques for large-scale graphs and address challenges like scalability and interpretability. The chapter also highlights real-world applications in social networks, bioinformatics, and recommendation systems, emphasizing the transformative potential of GNNs in various domains.

Chapter 2 explores the transformative potential of machine learning in revolutionizing neurology care, offering enhanced diagnosis, disease progression prediction, and personalized treatments. Successful integration of ML into clinical practice requires navigating a complex landscape of ethical, technological, and collaborative challenges as algorithms analyze increasingly sophisticated neurological data—ranging from neuroimaging to genetic markers and behavioral features. This chapter advocates for an interdisciplinary approach to address persistent issues such as data bias, patient privacy, the opacity of black-box systems, and the need for updated regulatory frameworks while emphasizing transparency and human oversight. The aim is to advance data quality, develop interpretable designs, and embed ethical considerations to fully realize ML's potential in neurology.

Chapter 3 examines the interplay between brain connectomics and graph theory in the mapping and analysis of neural networks. It emphasizes their significance in comprehending cognitive functioning and neurological illnesses, providing insights for clinical applications and individualized therapies.

Chapter 4 examines the role of GNNs in advancing the field of brain connectomics, focusing on how these networks can improve our understanding of brain connectivity and its disruptions in neurological disorders. Brain connectomics consists of structural, functional, and effective connectivity, which can be analyzed using GNNs. These networks provide a sophisticated framework to analyze complex, non-Euclidean data, enabling exploration of relationships between brain regions and their functions. The research demonstrates that GNNs offer superior performance in disease diagnosis and cognitive performance prediction. For instance, GNNs predicted Alzheimer's disease with 87% accuracy versus 79% for SVM. Similar improvements were seen for Parkinson's and schizophrenia. Challenges such as data quality, model interpretability, and computational scalability remain.

Chapter 5 explores the application of GNNs in classifying and diagnosing neurological diseases by leveraging their ability to model complex brain region relationships in neuroimaging data. Unlike traditional methods, GNNs capture non-Euclidean patterns, enhancing early disease detection, progression prediction, and personalized treatment. The chapter also discusses challenges such as data scarcity, computational demands, and model interpretability while highlighting future research directions.

Chapter 6 investigates how multimodal data GNNs could combine several kinds of data to address practical challenges. Many problems in healthcare, autonomous systems, and e-commerce require merging multiple data sources—text, images, and sensor measurements. Using their representation as a graph, GNNs provide a means of data processing. Key ideas of GNNs and several multimodal integration methods—including early, late, hybrid, attention, and tensor fusion—are introduced. Applications in social networking, self-driving cars, e-commerce, and healthcare are covered. The chapter emphasizes challenges such as data homogeneity, scalability, missing data, and interpretability, and suggests future research paths, including scalability improvements, interpretability advancements, and real-time applications.

In Chap. 7, GNNs have made strides in learning complex relationships between biological entities like genes, proteins, and metabolites for accurate biomarker discovery. GNNs are applied to gene expression and protein interaction data to identify key biomarkers of cancer and other diseases. Findings show that GNNs outperform traditional models: accuracy 92% vs. 75% (SVM) and 78% (Random Forest), with higher precision and recall. This suggests GNNs can enhance biomarker discovery and precision medicine.

Chapter 8 explores applications in cognitive neuroscience, focusing on computational and neuroimaging approaches to memory and learning. The chapter defines cognitive processes and examines how neuroscience techniques help unravel human cognition. Cutting-edge computational models and neuroimaging methods for studying memory and learning are discussed, highlighting applications that enhance understanding of brain function and bridge theoretical neuroscience with practical applications.

Chapter 9 explores how GNNs integrate with several modalities of multimodal neuroimaging data, including structural MRI (sMRI), functional MRI (fMRI), diffusion tensor imaging (DTI), genetic biomarkers, and clinical assessments. The framework improves disease classification accuracy and discovers individualized biomarkers through architectures like GCNs and GATs. It delivers robustness and flexibility across neurological conditions (Alzheimer's, Parkinson's, and epilepsy) and outperforms previous models (CNNs and RNNs) with an accuracy of 89%, precision of 87%, recall of 90%, and F1-score of 88%. GNN demonstrates the highest precision (90%) and recall (88%) for AD. This chapter contributes to computational neuroscience and AI-driven neurological diagnostics, benefiting researchers, clinicians, and data scientists.

Chapter 10 explores the use of GNNs in identifying potential calpain-10 inhibitors for neurological disorder therapy. Computational methods such as AlphaFold, DeepBindPoc, and DeepBindGCN provide insights into protein structure and ligand interactions, and AI is transforming drug development. The study aims to discover active binding sites, predict receptor structure, and screen therapeutic options using molecular dynamics and AI modeling, improving virtual screening accuracy. This chapter advances AI-assisted medication development and opens the door to novel treatments.

Chapter 11 delves into the foundational concepts of GNNs, highlighting their superiority over traditional non-graph-based approaches in handling multimodal and longitudinal clinical data. We explore potential applications in structuring patient networks, modeling disease progression, and uncovering biochemical interactions. Real-world applications illustrate how GNNs contribute to disease classification, patient subgroup identification, and drug discovery—accelerating drug repurposing and improving treatment strategies. Despite challenges in clinical adoption, the chapter provides insights to overcome them, ensuring GNNs can be effectively integrated into medical practice and contribute to precision medicine.

About the Book

Graph Neural Networks (GNNs) are at the forefront of altering neurological research and diagnostics in an era where artificial intelligence is revolutionizing medicine. The foundations of GNNs, their sophisticated structures, and their vital applications in comprehending brain illnesses are all thoroughly explored in this groundbreaking work. This book explores how GNNs perform better than conventional machine learning models when evaluating complex neuroimaging and multimodal data, with a focus on brain connectomics, illness classification, biomarker identification, and medication development. It showcases state-of-the-art research showing how GNNs improve Alzheimer's, Parkinson's, schizophrenia, and epilepsy predictions, providing fresh perspectives on early identification and individualized care.

This book addresses important issues including scalability, interpretability, and ethical considerations in AI-driven healthcare in addition to applications. It enables academics, physicians, and AI specialists to use GNNs for precision medicine and biomedical innovation by bridging the gap between computational neuroscience and clinical practice. This book is an essential resource for anybody working in the next frontier of neurological diagnostics and therapies powered by Graph Neural Networks, regardless of whether they are AI researcher, neuroscientist, or healthcare professional.

Contents

About the Editors and Contributors

About the Editors

Md. Mehedi Hassan is currently a Ph.D. researcher in STEM (Computer and Information Science) at the University of South Australia, working on a fully funded research project. He started his Ph.D. in 2025, building on a strong academic background in computer science and engineering. He holds an M.Sc. in Computer Science and Engineering from Khulna University, Bangladesh, completed in 2024, and a B.Sc. in Computer Science and Engineering from North Western University, completed in 2022. His research spans computer science engineering and data science, with a strong focus on predictive analysis and expert system development. He has authored 73 research papers and edited 5 books, actively contributing to the academic community. He serves as a peer reviewer for over 80 prestigious journals and collaborates extensively in interdisciplinary research. As a trainer for the VCourse platform, he has educated more than 250 students over the past two years, sharing knowledge in emerging technologies and research methodologies. Beyond publications, he has actively engaged in intellectual property development, with several patents filed and three already granted in his name. His current research interests include computational neuroscience, machine learning for healthcare, and predictive modeling for biometrics.

Anindya Nag obtained an M.Sc. in Computer Science and Engineering from Khulna University in Khulna, Bangladesh, and a B.Tech. in Computer Science and Engineering from Adamas University in Kolkata, India. He is currently a lecturer in the Department of Computer Science and Engineering at the Northern University of Business and Technology in Khulna, Khulna, Bangladesh. His research focuses on health informatics, medical Internet of Things, neuroscience, and machine learning. He serves as a reviewer for numerous prestigious journals and international conferences. He has authored and co-authored about 47 publications, including journal articles, conference papers, and book chapters, and has co-edited 9 books.

Shariful Islam is a physician trained in Cardiology with a PhD and postdoctorate fellowships in digital health. He is currently an NHMRC Emerging Leadership Fellow and Group Leader for Global Health at the Institute for Physical Activity and Nutrition at Deakin University and an Honorary Professor at the Noncommunicable Diseases Department, Bangladesh University of Health Sciences. Previously, he was an Honorary Senior Fellow at the George Institute for Global Health and an Honorary Senior Lecturer at the Sydney Medical School, University of Sydney. His research focuses on the design and conduct of large-scale epidemiological studies, meta-analyses, and clinical trials testing comparative effectiveness strategies for diabetes and cardiovascular diseases. In addition, his research also focuses on the use of mHealth, wearable devices, biosensors, and innovative information technologies for the prevention and management of chronic diseases.

Herat Joshi is a visionary leader in the field of Healthcare Informatics, healthcare technology, and data management, currently serving as the Data Science Team Lead at Great River Health Systems, USA. Holding a Ph.D. in Computer Science & Engineering, he has a distinguished track record of pioneering healthcare solutions that integrate AI, IoT, and high-performance computing to enhance clinical outcomes. His notable achievements include leading EHR implementations across multiple health systems, advancing interoperability, and receiving prestigious accolades such as the Outstanding Leadership Award at the Health 2.0 Conference. A recognized thought leader, he is a Fellow of the American College of Health Data Management (FACDM), Vice Chair of an American Medical Informatics Association (AMIA) Workgroup, Senior Member of IEEE, Vice Chair of IEEE Iowa-Illinois Section, and a Gartner Ambassador. He also contributes as a reviewer and editor for leading scientific journals.

Contributors

Shake Ibna Abir Department of Mathematics & Statistics, Arkansas State University, Jonesboro, AR, USA

Akash Adhikary Department of Computer Science and Engineering, Victorian Institute of Technology, Melbourne, VIC, Australia

Abdullah Al Abid Department of Software Engineering, Faculty of Science and Information Technology, Daffodil International University-Bangladesh, Daffodil Smart City, Birulia, Bangladesh

Md Rittique Alam Department of Computer Science, Faculty of Science and Information Technology, American International University-Bangladesh, Kuratoli, Bangladesh

Sarder Abdulla Al Shiam Department of Management, St. Francis College, New York, NY, USA

Thirunavukarasou Anand Faculty of Clinical Research, Sri Ramachandra Institute of Higher Education and Research, Chennai, Tamil Nadu, India

Dhanushkodi Balamurugan Department of BBA, Vels Institute of Science, Technology and Advanced Studies, Chennai, Tamil Nadu, India

Teresa Jency Bala Computer Science and Engineering Discipline, Khulna University, Khulna, Bangladesh

Rezuana Haque Department of Computer Science and Engineering, BGC Trust University Bangladesh, Chittagong, Bangladesh

Mahade Hasan School of Software, Nanjing University of Information Science and Technology, Nanjing, China

Md. Mehedi Hassan Computer Science and Engineering, Khulna University, Khulna, Bangladesh
Computer Science and Engineering, Khulna University, Khulna, Bangladesh

Monirul Islam Department of Management, Sister Nivedita University, Kolkata, West Bengal, India

Nazrul Islam Khan Department of Mathematics & Statistics, Stephen F. Austin State University, Nacogdoches, TX, USA

Hirak Mondal Computer Science and Engineering Discipline, Khulna University, Khulna, Bangladesh

Anindya Nag Department of Computer Science and Engineering, Northern University of Business and Technology Khulna, Khulna, Bangladesh

Bernard-Marie Onzo School of Computer Science, Nanjing University of Information Science and Technology, Nanjing, China

Konda Mani Saravanan B Aatral Biosciences Private Limited, Bangalore, Karnataka, India

Renganathan Senthil Department of Bioinformatics, Vels Institute of Science, Technology and Advanced Studies, Chennai, Tamil Nadu, India

Shaharina Shoha Department of Mathematics & Statistics, Arkansas State University, Jonesboro, AR, USA

Yu Xue School of Computer Science, Nanjing University of Information Science and Technology, Nanjing, China
School of Software, Nanjing University of Information Science and Technology, Nanjing, China

Farhana Yasmin School of Computer Science, Nanjing University of Information Science and Technology, Nanjing, China
School of Software, Nanjing University of Information Science and Technology, Nanjing, China

Chapter 1
Graph Neural Network Architectures and Algorithms

Farhana Yasmin, Mahade Hasan, Yu Xue, Md. Mehedi Hassan, and Bernard-Marie Onzo

1.1 Introduction

As graphs are a general data structure that may be utilized to illustrate relationships between entities, they are a good choice for issues involving interconnected systems. From social networks and molecular structures to transportation systems and knowledge graphs, graph data is pervasive in natural and artificial systems [1]. Understanding and analyzing this data type requires specialized techniques beyond traditional machine-learning methods designed for grid-structured data.

GNNs are a powerful class of models that leverage relationships between nodes and their neighbors to learn from graph-structured data. Unlike traditional neural networks, GNNs are particularly effective at capturing the underlying topology and contextual information of graphs, making them highly adaptable for various applications [2]. Their capability to encode features at the graph, node, and edge levels has transformed domains such as social network analysis, bioinformatics, and recommendation systems. Despite their success, GNNs face several challenges, including scalability for large graphs, over-smoothing of features in deeper architectures, and difficulties in training on dynamic or heterogeneous graphs [3]. Additionally, enhancing the interpretability and explainability of GNNs remains an ongoing

F. Yasmin (✉) · Y. Xue · B.-M. Onzo
School of Computer Science, Nanjing University of Information Science and Technology, Nanjing, China
e-mail: farhana@nuist.edu.cn; xueyu@nuist.edu.cn

M. Hasan
School of Software, Nanjing University of Information Science and Technology, Nanjing, China
e-mail: mhasan@nuist.edu.cn

M. M. Hassan
Computer Science and Engineering, Khulna University, Khulna, Bangladesh

M. M. Hassan et al. (eds.), *Graph Neural Networks for Neurological Disorders*,
https://doi.org/10.1007/978-3-032-04315-3_1

research focus. Overcoming these challenges is essential for furthering the development and practical deployment of GNNs in real-world applications.

This chapter will explore the foundational concepts, architectures, and algorithms that underpin GNNs, providing a comprehensive overview of their capabilities and limitations. It will also delve into advanced techniques, highlight key challenges in graph-based learning, and discuss future directions to inspire continued innovation in this rapidly evolving field.

1.1.1 Overview of Graph Neural Networks (GNNs)

Graph-structured data, in which things are represented as nodes and relationships between them as edges, is processed and analyzed by GNNs, a specific class of deep learning systems. GNNs are naturally adapted for irregular, non-Euclidean data, in contrast to typical neural networks that function on sequential data, like text, or grid-like structures, like photos. This capability makes them highly effective for problems where the relational structure between entities is as important as the features of the individual entities themselves.

Message passing is a fundamental component of GNNs that allows nodes to iteratively exchange information with their neighbors. A node modifies its representation by combining data from its close neighbors in each cycle [4]. This iterative process keeps going until a meaningful representation that encodes the graph's local and global features is produced. After that, these representations can be applied to a number of tasks, including node classification, which aims to predict a node's category; link prediction, which entails identifying missing or possible edges in the graph; and graph classification, which assigns entire graphs to particular classes [5]. Figure 1.1 illustrates the graph-level learning of GNN.

The capacity of GNNs to generalize across different kinds of graphs and datasets is one of its main advantages. For instance, in social networks, GNNs can identify patterns such as community structures or influence dynamics by analyzing the relationships between users [6]. In bioinformatics and drug discovery, they are used to predict molecular properties, analyze protein-protein interactions, and design new compounds. Similarly, in recommendation systems, GNNs effectively model user-item interactions to generate personalized recommendations [7]. Their ability to integrate graph topology with node and edge features makes them a powerful tool for tasks requiring relational reasoning.

GNNs have proven to be remarkably effective at resolving intricate graph-based issues in a variety of domains, expanding the possibilities of machine learning on graph data. By effectively capturing the structural and contextual information present in graphs, GNNs have opened new avenues for solving challenges that were previously considered difficult or even intractable, establishing themselves as a cornerstone in graph-based machine learning.

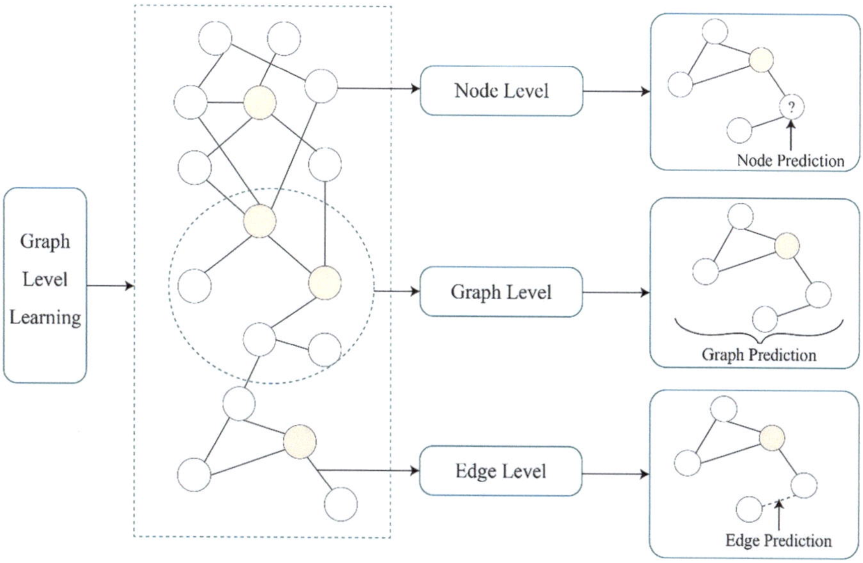

Fig. 1.1 Illustrates the graph-level learning of GNN

1.1.2 Applications of GNNs in Real-World Problems

GNNs have shown immense potential in addressing a variety of real-world problems across diverse domains, thanks to their ability to model complex relationships and dependencies in graph-structured data. By leveraging the power of message passing and representation learning, GNNs can extract meaningful insights from data with relational and structural components, making them a versatile tool for solving practical challenges.

One prominent application of GNNs is in social network analysis, where they are used to model user interactions, detect communities, and predict social ties. GNNs can also identify influential nodes in social graphs and analyze the spread of information or misinformation. Similarly, in bioinformatics and chemistry, GNNs have revolutionized tasks such as protein structure prediction, drug discovery, and molecular property prediction by modeling molecular graphs and protein-protein interaction networks [8]. These advancements are crucial for understanding biological systems and accelerating the discovery of new treatments.

Another key area where GNNs excel is recommendation systems, where they analyze user-item interaction graphs to provide personalized recommendations [9]. By capturing the underlying structure of user preferences and item similarities, GNNs outperform traditional collaborative filtering methods. In transportation and urban planning, GNNs are used to analyze road networks, optimize traffic flow, and predict transportation demands [10]. They also play a significant role in knowledge graphs, where they help in tasks such as relation extraction, entity linking, and graph-based reasoning, enabling advancements in natural language processing and information retrieval.

In cybersecurity, GNNs have been applied to model network traffic and detect anomalies, such as intrusions or fraud, by analyzing the relationships between entities like devices, users, or transactions. In finance, they are used for fraud detection, credit scoring, and stock market prediction by modeling interactions within transaction networks or market dynamics [11]. Furthermore, GNNs are finding applications in physics and materials science, where they are used to model physical systems, predict material properties, and simulate complex interactions.

GNNs have become an essential tool for tackling a wide range of real-world problems, demonstrating their adaptability and effectiveness in diverse domains. Their ability to integrate structural and contextual information enables them to address challenges that traditional machine-learning methods struggle to solve, paving the way for innovative solutions and transformative advancements in various fields.

1.1.3 Key Challenges in Graph-Based Learning

While GNNs have achieved significant success in processing and analyzing graph-structured data, they face several challenges that hinder their scalability, efficiency, and applicability in complex real-world scenarios. Addressing these challenges is crucial for further advancing the field of graph-based learning and expanding its practical utility.

The primary challenge is scalability. In real-world applications such as knowledge graphs and social networks, graphs frequently have millions or even billions of nodes and edges. Processing such large-scale graphs requires substantial computational resources and memory, making it difficult to train GNNs efficiently [12]. Techniques such as sampling, graph partitioning, and distributed computing have been developed to address this issue, but they come with trade-offs in terms of accuracy and implementation complexity.

Another critical issue is over-smoothing, a phenomenon where the representations of nodes in a graph become indistinguishable as the number of GNN layers increases. This leads to a loss of meaningful information, especially in tasks requiring detailed local structure [13]. Similarly, over-squashing is a challenge in graphs with high-degree nodes, where information from distant nodes is compressed into limited representation space, resulting in poor information flow.

Handling dynamic and heterogeneous graphs is another major challenge. Many real-world graphs evolve or contain multiple types of nodes and edges with distinct relationships. Adapting GNNs to efficiently capture temporal dynamics and heterogeneity requires specialized architectures, which often increase the complexity of the model [14]. Furthermore, sparsity in graph data can lead to challenges in learning meaningful patterns, especially when the graph has nodes or edges with limited information.

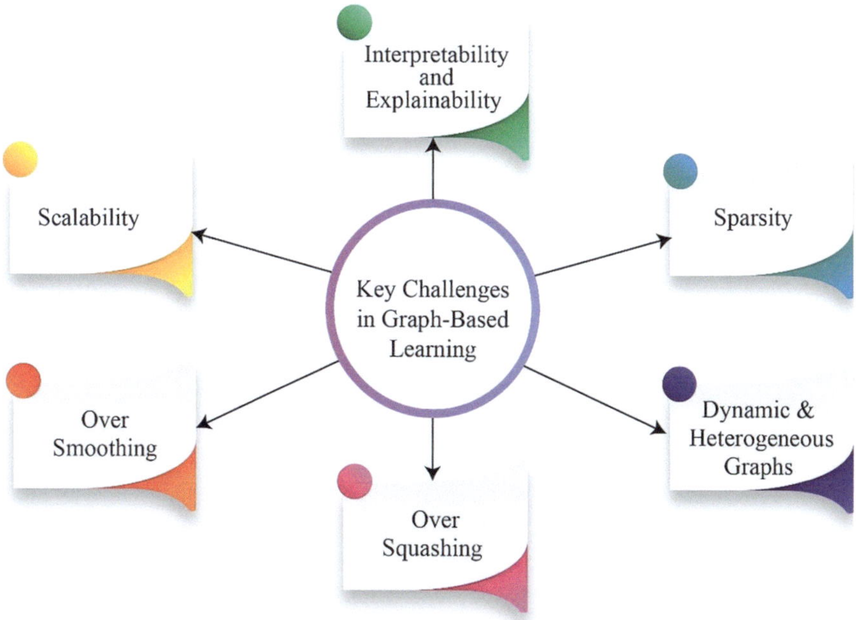

Fig. 1.2 Illustrates the key challenges in graph-based learning

Finally, interpretability and explainability remain significant concerns in GNNs. Understanding how GNNs make decisions and which parts of the graph contribute most to their predictions is essential, particularly in critical applications such as healthcare and finance [15]. However, the intricate message-passing mechanisms and complex graph structures make it challenging to provide clear explanations for the predictions of GNNs. Figure 1.2 illustrates the key challenges in graph-based learning.

Overcoming these challenges will require innovative methods and robust techniques that balance scalability, accuracy, and interpretability. By addressing these limitations, the potential of GNNs can be fully realized, enabling their widespread application in increasingly complex and large-scale graph-based problems.

1.2 Fundamentals of Graph Neural Networks

GNNs are specialized models designed to process graph-structured data. In this section, we will introduce the key concepts of graph theory and the basics of GNNs, focusing on graph components, representations, and the message-passing mechanism that underpins GNNs.

The fundamental elements of graphs, including nodes, edges, and adjacency matrices, will be addressed, along with the many kinds of graphs, including directed, undirected, and bipartite graphs. We will also go over how GNNs process graph data using graph representations such as feature matrices and embeddings. We will then examine the message-passing method, in which nodes update their feature representations by combining data from their neighbors. Lastly, we will examine how GNNs produce representations at the node, edge, and graph levels in order to carry out tasks like graph and node classification. Understanding how GNNs work is made easier by this part, which also prepares the reader for a deeper examination of advanced architectures and algorithms.

1.2.1 Graph Theory Foundations

The mathematical study of graphs, which are structures used to represent pairwise relations between things, is known as graph theory. Understanding the fundamental components of graphs, such as nodes (vertices) and edges (connections), is essential for grasping how GNNs leverage graph structures for learning. In this section, we will explore the basic building blocks of graphs and how they are represented mathematically, which form the foundation for processing data in GNNs.

Nodes, which stand for entities or objects, and edges, which represent connections or relationships between nodes, make up a graph. Depending on whether the interaction between nodes has a specified direction (e.g., a one-way street in a road network), these edges can be either directed or undirected. Additionally, edges can be weighted, meaning each edge has a value or cost associated with it, or they can be unweighted, where all edges are treated equally [16]. An adjacency matrix, a square matrix with rows and columns representing nodes and values denoting the existence or weight of an edge connecting nodes, is frequently used to depict graphs. For instance, a1 denotes the presence of an edge in a binary adjacency matrix, whereas a0 denotes its absence. The matrix entries in a weighted graph indicate the cost or weight assigned to each edge. The feature matrix, in which every node has a vector of characteristics that characterize its attributes, is another popular representation. Bipartite graphs, which are composed of two different sets of nodes with edges only connecting nodes from different sets, are one of the many different types of graphs [17]. This type of graph is useful for modeling relationships like those found in recommendation systems, where one set may represent users and the other set represents items. Figure 1.3 illustrates different types of graphs with fundamental components of graphs.

Therefore, understanding the fundamental components of graphs and how they are represented is crucial for working with GNNs. This foundation allows us to move forward with exploring how these structures are utilized in GNNs to process and learn from graph data.

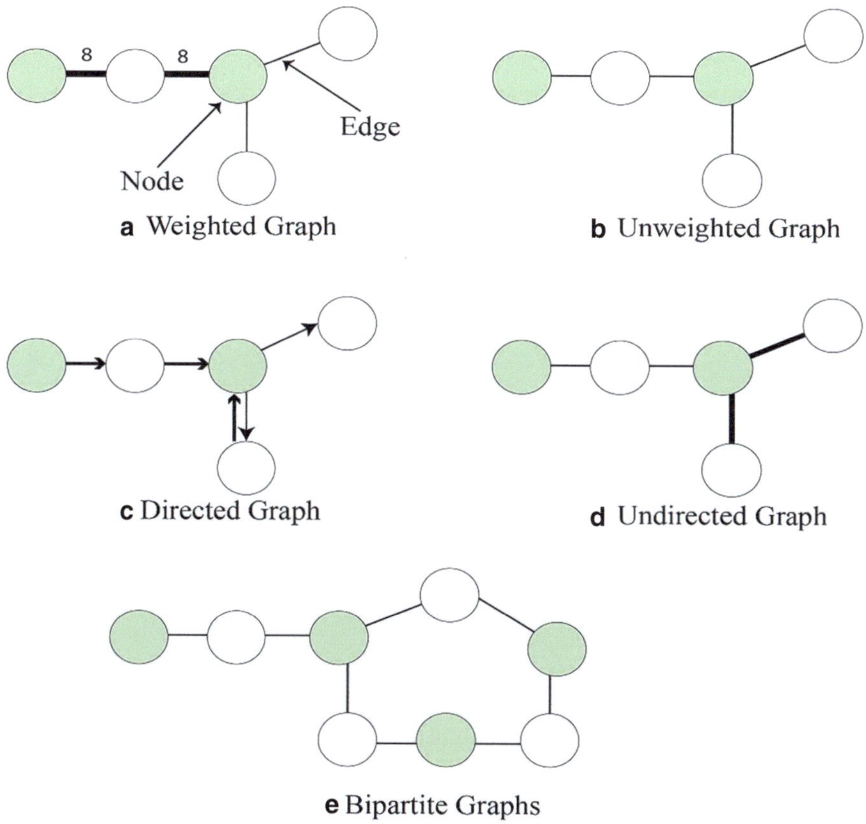

Fig. 1.3 Illustrates different types of graphs with fundamental components of graphs

1.2.2 Message-Passing Mechanism in GNNs

The message-passing mechanism is at the foundation of GNNs, enabling them to collect and propagate information across a graph. This mechanism allows nodes to share information with their neighbors, which is then used to update their feature representations. In this section, we will explore how the message-passing process works and why it is crucial for the performance of GNNs.

Each node collects and mixes its own features with those of its neighbors during the message transit process. This aggregation step is typically done through operations like summation, averaging, or maximum pooling. After the aggregation, the node updates its feature representation by applying a transformation, often involving a neural network layer (e.g., a fully connected layer or a nonlinear activation function). This process is repeated across multiple iterations (or layers), allowing nodes to capture information from increasingly distant neighbors [18]. Figure 1.4 illustrates the message-passing techniques in GNN.

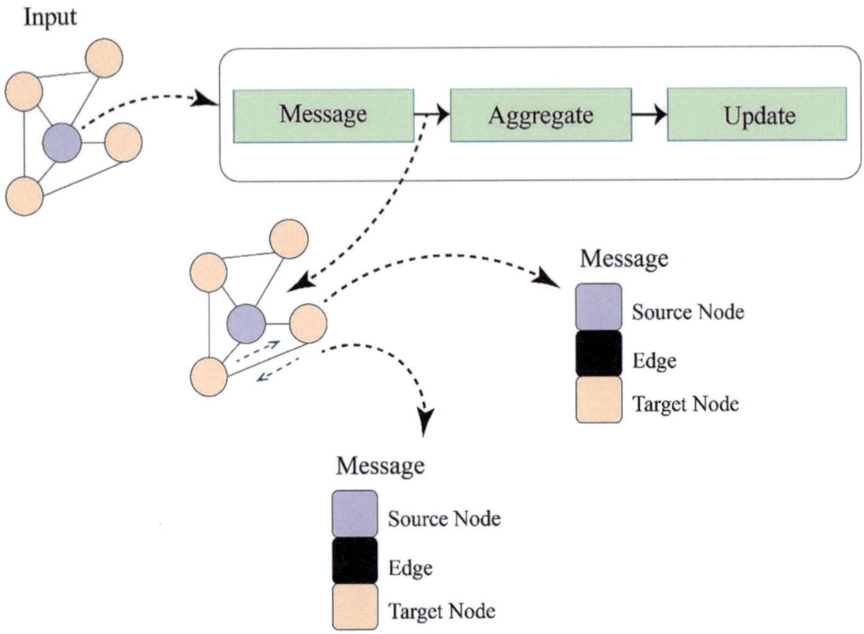

Fig. 1.4 Illustrates the message-passing techniques in GNN

The key aspect of message passing is that it allows GNNs to learn from the graph structure itself. Unlike traditional neural networks, where data is passed through layers sequentially, GNNs use the structure of the graph to define how information flows between nodes. For tasks like node classification, link prediction, and graph classification, GNNs are very effective because of this process, which allows them to recognize both local and global patterns inside the graph [19].

The message-passing mechanism in GNNs is fundamental to their ability to process graph-structured data. By enabling nodes to aggregate and propagate information across the graph, GNNs can learn rich representations that reflect the underlying structure of the graph. This process forms the basis for more advanced GNN architectures and applications.

1.2.3 Node, Edge, and Graph-Level Representations

In GNNs, learning effective representations at the node, edge, and graph levels is crucial for solving tasks that depend on graph-structured data. GNNs are able to capture complex dependencies and relationships within the graph due to these hierarchical representations. In this section, we will explore how GNNs extract features at different levels and compare how they differ from traditional neural network architectures like CNNs and RNNs.

At the node level, GNNs learn feature representations for individual nodes by aggregating information from their neighbors. These node-level representations capture local structural patterns, which are important for tasks like node classification or link prediction [20]. At the edge level, GNNs can also learn edge representations, which are useful when predicting relationships or interactions between nodes. These representations are typically learned by aggregating features from the nodes that are connected by the edge. At the graph level, GNNs generate a holistic representation of the entire graph, which can be used for tasks like graph classification or graph generation. The hierarchical nature of GNNs, where node-level representations are combined to form edge-level and graph-level representations, enables the model to capture both local and global patterns in the graph structure [21]. Figure 1.5 illustrates the architecture of CNN, RNN, and GNN.

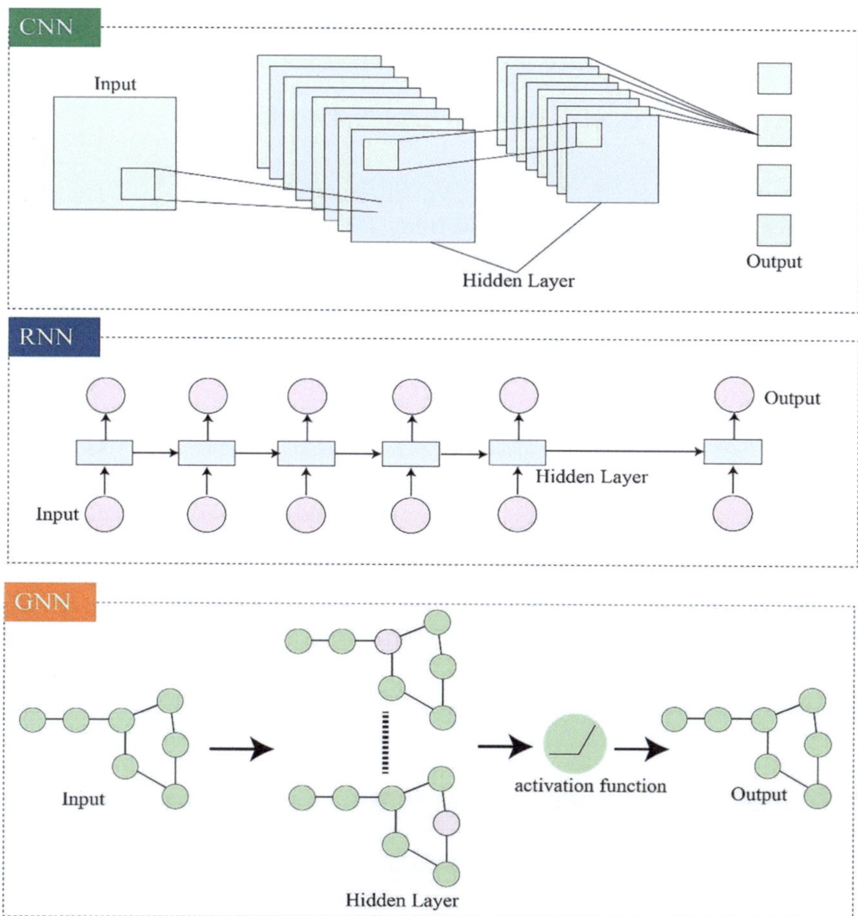

Fig. 1.5 Illustrates the architecture of CNN, RNN, and GNN

Graph-structured data handling is one of the main advantages of GNNs over more conventional neural networks, including CNNs and RNNs. GNNs are specifically made to learn from the irregular structures of graphs, where relationships between elements are not defined by a fixed order or grid [24]. CNNs are particularly good at processing grid-like data (like images) [22], and RNNs are good at processing sequential data (like time series) [23]. This flexibility allows GNNs to generalize across a wide variety of tasks and domains that involve graph data, such as social network analysis, recommendation systems, and molecular biology.

Therefore, GNNs are effective for graph-based learning problems due to their capacity to learn hierarchical representations at the node, edge, and graph levels. The capacity of GNNs to simulate intricate dependencies in graph-structured data distinguishes them from classic neural networks and creates new opportunities for applications that were previously challenging to handle with traditional designs.

1.3 Architectures of Graph Neural Networks

Many architectures of GNNs are suited to different kinds of graph-related tasks, making them an effective tool for learning from graph-structured data. These architectures exploit different aspects of graph data, such as node relationships, edge connectivity, and graph structures, to improve performance in tasks like classification, link prediction, and clustering. In this section, we will explore some of the key architectures of GNNs and the underlying principles that make them effective.

One of the foundational architectures of GNNs is the Graph Convolutional Network (GCN), which uses convolutional layers designed specifically for graph-structured data. GAT introduces an attention mechanism to dynamically assign different weights to node neighbors based on their importance. For sequential graph data, Graph RNNs utilize recurrent architectures to capture temporal dependencies. Graph Autoencoders provide another approach by learning graph representations through an unsupervised autoencoding process. Finally, the distinction between Spatial and Spectral architectures is an important one, as these methods differ in how they approach graph convolutions, either directly on node neighborhoods (spatial) or in the graph's spectral domain (spectral).

Each of these architectures presents unique advantages depending on the type of graph data and task at hand. Understanding their workings will help us apply them more effectively to real-world issues like recommender systems, bioinformatics, and social network analysis. In the upcoming subsections, we will examine each of these architectures in more detail to highlight their unique processes and applications.

1.3.1 Graph Convolutional Networks (GCN)

GCNs are a powerful class of GNNs designed to operate on graph-structured data. Unlike traditional CNNs that work with grid-like structures such as images, GCNs perform convolution operations on graphs, where the relationships between data points are captured by nodes and edges. For applications where the data is represented as graphs, such as node classification, graph classification, and link prediction, GCNs are especially well-suited.

At the heart of GCNs is the graph convolutional layer, which aggregates information from a node's neighbors to compute its updated feature representation [25]. The operation relies on the graph structure, typically represented as an adjacency matrix. Specifically, for a given graph with N nodes, the feature matrix $X \in R^{N \times F}$ (where F is the number of features per node) is updated iteratively by aggregating information from neighboring nodes, weighted by their connections. Mathematically, a single graph convolution operation in a GCN can be defined as Eq. (1.1):

$$H(1+1) = \sigma\left(\hat{A}H(1)W(1)\right), \tag{1.1}$$

where:

- $H^{(l)}$ is the matrix of node representations at layer l, with $H^{(0)} = X$ (initial node features).
- $\hat{A} = A + I$ is the normalized adjacency matrix, where A is the adjacency matrix of the graph (with entries $A_{ij} = 1$ if node ii is connected to node j and 0 otherwise), and I is the identity matrix to ensure that each node also aggregates information from itself.
- $W^{(l)}$ is the learned weight matrix for layer l.
- σ is a non-linear activation function, typically ReLU.

This operation aggregates information from a node's neighbors and applies a learned transformation through $W^{(l)}$ to compute the updated feature representation. The process is repeated for multiple layers, allowing nodes to progressively incorporate information from increasingly distant neighbors.

A crucial aspect of GCNs is the graph convolution operation's reliance on spectral graph theory. In the spectral domain, the graph convolution is related to filtering signals defined on the graph using the graph's Laplacian eigenbasis. The convolution operation essentially filters these signals, allowing for the extraction of meaningful patterns from the graph structure. In practice, the graph convolution operation can be interpreted as a message-passing scheme where each node aggregates messages (features) from its neighbors, which are then passed through a learned transformation [26]. This mechanism allows GCNs to effectively capture both local (local node neighborhoods) and global (graph-wide) dependencies, making them suitable for a wide variety of graph-based tasks. Algorithm 1.1 demonstrates the typical GCN techniques.

Algorithm 1.1 Graph Convolutional Networks (GCNs)

1: Step 1: **Initialization**
2: Initialize feature matrix X; adjacency matrix A;
3: Initialize weight matrix $W^{(l)}$ for each layer l; bias vectors $b^{(l)}$ for each layer l;
4: Initialize learning rate η; number of epochs K; optimizer (e.g., Adam);
5: Step 2: **Train Model**
6: **for** epoch in range(K) do:
7: Step 3: **Graph Convolution Layer Calculation**
8: Compute node embeddings $H^{(l)} = \sigma(\hat{A}\ H^{(l-1)}W^{(l)} + b^{(l)})$
9: Repeat the above operation for all layers until $H^{(L)}$, the final node representations, are computed.
10: Step 4: **Loss Calculation**
11: Compute the loss function L (e.g., cross-entropy loss for classification, MSE for regression tasks).
12: Step 5: **Optimization**
13: Backpropagate gradients ∇L
14: Update parameters $W^{(l)}$, $b^{(l)}$ using optimizer (e.g., Adam).
15: **end for**
16: Step 6: **Output**
17: After K epochs, return the final node representations $H^{(L)}$

In summary, GCNs provide a robust method for learning graph-structured data by applying convolutional operations that aggregate information from a node's neighbors. The key mathematical formulation leverages a spectral graph approach to effectively learn node representations that incorporate both local and global graph features. GCNs have shown their ability to describe intricate network relationships by doing well in tasks including node classification, link prediction, and graph classification.

1.3.2 Graph Attention Networks (GAT)

GATs introduce the concept of attention mechanisms to GNN, allowing nodes to selectively focus on the most relevant neighbors during the aggregation process. While Graph Convolutional Networks (GCNs) rely on a predefined adjacency matrix to aggregate features from neighboring nodes, GATs dynamically adjust the importance of each neighbor by computing an attention coefficient for each connection. This enables GATs to perform more sophisticated and flexible neighborhood aggregation, where nodes can weigh the influence of different neighbors based on their relevance.

The key feature of GATs is the self-attention mechanism, which calculates attention coefficients for each pair of nodes based on their features and the structure of the graph. These attention coefficients are then used to weigh the feature aggregation process, allowing nodes to focus on more important neighbors while down-weighting the influence of irrelevant or noisy neighbors [27]. The attention mechanism in GATs is computed via the scaled dot-product attention, which has been widely used in models like transformers. Mathematically, the graph attention layer in a GAT can be defined as Eqs. (1.2) and (1.3):

$$\alpha_{ij} = \text{soft}\max_j\left(\text{LeakyReLU}\left(a^T\left[Wh_i - Wh_j\right]\right)\right) \quad (1.2)$$

$$h'_i = \sum_{j \in N(i) \cup \{i\}} \alpha_{ij} Wh_j, \quad (1.3)$$

where:

- $h_i$and h_j are the feature vectors of nodes i and j, respectively.
- W is a learnable weight matrix applied to the node features.
- || parallel denotes concatenation of the node feature vectors.
- *a* is a learnable attention vector used to compute the attention score between nodes i and j.
- α_{ij} is the attention coefficient that determines the weight of node *j*'s contribution to node *i*'s updated feature representation.
- *N*(i) is the set of neighbors of node i, and the sum aggregates the weighted features of all neighbors.

The attention mechanism in GATs allows the model to adaptively determine how much influence each neighbor should have based on the similarity between the feature representations of the nodes. This is particularly useful in scenarios where the relationships between nodes are not uniform, and some neighbors might have more relevant information for a given task than others. One of the key advantages of GATs is their ability to handle heterogeneous graphs with varying node and edge characteristics. The attention mechanism allows for a flexible and fine-grained weighting of neighbors, which is especially useful when dealing with graphs where nodes have varying degrees of importance or when the graph is sparse [28]. Algorithm 1.2 demonstrates the typical GAT techniques.

Algorithm 1.2 Graph Attention Networks (GATs)
1: Step 1: **Initialization**
2: Initialize feature matrix X; adjacency matrix A;
3: Initialize attention weight matrix $W^{(l)}$ for each layer l; bias vectors $b^{(l)}$ for each layer l;
4: Initialize learnable attention parameters α_{ij} for each pair of connected nodes;
5: Initialize learning rate η; number of epochs K; optimizer (e.g., Adam);
6: Step 2: **Train Model**
7: **for** epoch in range(K) do:
8: Step 3: **Graph Attention Layer Calculation**
9: **for** each layer l, compute the attention coefficients α_{ij} for each edge (i,j) using the self-attention mechanism:

10: Compute the updated node embeddings by aggregating the features of each node's neighbors weighted by the attention coefficients:
11: Repeat this operation for all layers to compute the final node representations $H^{(L)}$
12: Step 4: **Loss Calculation**
13: Compute the loss function L (e.g., cross-entropy loss for classification, MSE for regression tasks).
14: Step 5: **Optimization**
15: Backpropagate gradients ∇L
16: Update parameters $W^{(l)}$, $b^{(l)}$, and attention weights α_{ij} using optimizer (e.g., Adam).
17: **end for**
18: **end for**
19: Step 6: **Output**
20: After K epochs, return the final node embeddings $H^{(L)}$.

Therefore, GATs enhance the traditional graph convolutional framework by introducing an attention mechanism that dynamically weighs the influence of neighbors. This flexibility allows GATs to focus on the most relevant information, making them more efficient and effective in graph-based tasks where relationships between nodes are complex and heterogeneous. GATs have shown significant improvements in tasks such as node classification, link prediction, and graph generation.

1.3.3 Graph Recurrent Neural Networks (Graph RNNs)

Graph Recurrent Neural Networks (Graph RNNs) extend the traditional recurrent neural networks (RNNs) to graph-structured data. While RNNs are designed to handle sequential data by maintaining a hidden state that captures the temporal dependencies between elements in a sequence, Graph RNNs aim to model sequential dependencies in graph-structured data by maintaining a hidden state for each node and updating it over time. This makes Graph RNNs well-suited for applications that involve dynamic graphs, where the relationships between nodes evolve, such as social network analysis, dynamic traffic prediction, and temporal link prediction.

The key idea behind Graph RNNs is to apply recurrent updates to the hidden states of the nodes, where the updates depend not only on the current node's features but also on the features of its neighbors. The recurrent process allows the network to capture temporal dependencies and changes in graph structure over multiple iterations [29]. These models combine the power of both graph-based learning and sequential processing to learn rich representations of nodes and edges over time. Mathematically, the update rule for a Graph RNN can be expressed as Eq. (1.4):

$$h_i^{(t+1)} = \sigma\left(W h_i^{(t)} + \sum_{j \in N(i)} \alpha_{ij} h_j^{(t)} + b \right), \tag{1.4}$$

where:

- $h_i^{(t)}$ is the hidden state of node i at time step t.
- b is a bias term.

The update rule shows how the hidden state of each node is updated at each time step by aggregating information from its neighbors (represented by the summation term) and using a recurrence mechanism to refine its state. One of the key advantages of Graph RNNs is their ability to model temporal dynamics in graphs, where the relationships between nodes are not static but evolve [30]. For example, in social networks, the interactions between users (nodes) change dynamically, and Graph RNNs can capture these changes by recurrently updating the states of the nodes based on the latest information from both the nodes themselves and their neighbors. Graph RNNs can be further extended to include more advanced mechanisms, such as attention and gated updates (e.g., in the case of Graph GRUs or Graph LSTMs), to allow for more sophisticated temporal dependencies and better handling of varying graph structures [31].

In summary, Graph RNNs are a powerful class of models that combine the sequential nature of RNNs with the structural dependencies inherent in graph data. By maintaining and updating node representations, Graph RNNs can effectively capture temporal dynamics in evolving graphs, making them particularly useful for tasks that require the modeling of time-varying relationships, such as temporal link prediction, dynamic graph generation, and forecasting in dynamic systems.

1.3.4 Graph Autoencoders and Variants

Graph Autoencoders (GAEs) are unsupervised learning models designed to learn efficient low-dimensional embeddings of graph-structured data. The core idea behind GAEs is like that of traditional autoencoders used for vector data: they attempt to encode graph data into a compact latent representation and then decode it back to reconstruct the original graph or its structural properties. This process enables the network to capture the inherent patterns and relationships within the graph structure, making GAEs highly suitable for tasks like link prediction, node classification, and graph generation.

1.3.4.1 Key Components of Graph Autoencoder

An encoder and a decoder are the two primary parts of a conventional graph auto-encoder.

(i) Encoder: The encoder transforms the graph data (usually represented by an adjacency matrix and node features) into a compact latent space. In the context of Graph Autoencoders, the encoder typically uses GCNs or GATs to aggregate information from neighboring nodes and learn the node embeddings.

(ii) Decoder: The Decoder reconstructs the graph or the relationships between nodes from the latent representations. For example, in link prediction, the decoder predicts the likelihood of an edge existing between two nodes based on their latent embeddings [32].

1.3.4.2 Summarized Process of Graph Autoencoder

The process can be summarized in the following steps:

(i) Encoding: A GCN or GAT-based encoder transforms the input graph (with its adjacency matrix and node features) into a low-dimensional latent space. The encoder typically operates on the graph structure and node features to generate embeddings for each node.
(ii) Latent Representation: Each node is represented by a latent vector, which captures the essential information about that node and its neighborhood.
(iii) Decoding: The latent representations are then passed through a decoder, which reconstructs the graph or predicts missing edges between nodes [33].

1.3.4.3 Variants of Graph Autoencoders

(i) Variational Graph Autoencoders (VGAE): One of the popular extensions of Graph Autoencoders is the Variational Graph Autoencoder (VGAE), which introduces a probabilistic interpretation of the encoding process. VGAE assumes that the latent space follows a specific distribution (usually Gaussian), and the encoder outputs parameters of this distribution (mean and variance) rather than deterministic embeddings. This variant introduces variational inference to optimize the model, making it more flexible and capable of handling uncertainty in the graph data.
(ii) Graph Convolutional Autoencoders (GCAE): This is a variant of the traditional Graph Autoencoder that uses graph convolutional layers in both the encoder and decoder. The goal is to use GCNs to learn both the node embeddings and the graph structure effectively. In this variant, the decoder might employ a graph convolutional network for reconstruction, enabling it to capture more complex relationships between nodes.
(iii) Adversarial Graph Autoencoders (AGAE): Another extension is the Adversarial Graph Autoencoder (AGAE), which introduces an adversarial training strategy like Generative Adversarial Networks (GANs). In AGAEs, a discriminator is added to the architecture to distinguish between real and fake node embeddings, guiding the encoder to generate more realistic graph representations.
(iv) Deep Graph Infomax (DGI): Deep Graph Infomax is another variant that uses contrastive learning to maximize mutual information between node representations and their surrounding context, providing a way to learn graph embeddings in an unsupervised manner. DGI can be considered a variant of Graph

Autoencoders with a focus on learning node-level representations while preserving global graph structure [34].

1.3.4.4 Applications of Graph Autoencoders

Graph Autoencoders have numerous applications in areas where learning graph representations is essential:

(i) Link Prediction: Predicting the likelihood of edges appearing in a graph, such as recommending new connections in social networks or identifying missing interactions in biological networks.
(ii) Graph Generation: Generating new graphs that follow the distribution of a given set of graphs, useful in areas such as molecular chemistry or network design.
(iii) Node Classification: Using learned representations to classify nodes in tasks like fraud detection or protein function prediction [35].

Algorithm 1.3 Graph Autoencoder (GAE)

Step 1: **Initialization**
Initialize feature matrix X; adjacency matrix A;
Initialize encoder weights W_{enc}, biases b_{enc};
Initialize decoder weights W_{dec}, biases b_{dec};
Initialize learning rate η; number of epochs K; optimizer (e.g., Adam);
Step 2: **Train Model**
for epoch in range(K) do:
Step 3: **Encoder**
Compute node embeddings $H = \text{Encoder}(X, A, W_{enc}, b_{enc})$
Step 4: **Decoder**
Reconstruct adjacency matrix $A_{hat} = \text{Decoder}(H, W_{enc}, b_{enc})$
Step 5: **Loss Calculation**
Compute loss $L_{recon} = \text{Loss}(A, A_{hat})$ (e.g., Mean Squared Error or Binary Cross-Entropy)
Step 6: **Optimization**
Backpropagate gradients: ∇L_{recon}
Update parameters using optimizer (e.g., Adam)
end for
Step 7: **Output**
Return the learned node embeddings H.

Algorithm 1.3 demonstrates the typical GAE techniques. GAEs and their variants provide powerful tools for unsupervised graph representation learning. By encoding the graph structure into low-dimensional embeddings and reconstructing the graph or its properties, they can capture the inherent patterns within graphs. Variants like VGAE, GCAE, AGAE, and DGI introduce enhancements such as probabilistic learning, adversarial training, and contrastive learning, further extending the applicability of Graph Autoencoders to a wide range of graph-based tasks.

1.3.5 Spatial vs. Spectral Architectures

In GNNs, the architecture design can broadly be classified into two categories: spatial and spectral approaches. Both strategies aim to learn representations of graph-structured data, but they operate differently in how they aggregate and propagate information across nodes.

Spatial GNNs focus on local neighborhoods for feature aggregation. In this approach, each node aggregates features from its immediate neighbors in the graph. This aggregation typically involves some form of pooling or summation operation, where a node's feature is updated based on the features of its neighboring nodes. The strength of spatial GNNs lies in their ability to learn from local structures, making them well-suited for tasks where the relationship between directly connected nodes is more important than global graph structure [36]. A common spatial architecture is the Graph Convolutional Network (GCN), where the feature of each node is updated by aggregating features from its neighbors through a weighted summation. On the other hand, spectral GNNs are based on the spectral properties of the graph, primarily focusing on the graph's Laplacian. In spectral methods, the graph is transformed into a frequency domain where the graph's structure is represented by its eigenvalues and eigenvectors. The graph convolution operation in this domain allows the model to consider global graph structure, enabling the propagation of node information across the entire graph, even between distant nodes. Spectral methods, such as Graph Fourier Transform (GFT), typically involve multiplying the graph signal with a filter that operates on the spectral components, making this approach sensitive to the global properties of the graph [37].

While both approaches have their merits, spatial architectures are more computationally efficient and are generally easier to implement. They operate on the node and edge features directly and can scale better to large graphs with dynamic structures. Spectral architectures, however, can capture more global information and relationships within the graph, but they are often computationally expensive and require the graph to be fixed and stationary, as they rely heavily on the eigendecomposition of the Laplacian, which can be costly for large graphs [38].

Therefore, the choice between spatial and spectral approaches depends on the specific graph task at hand. Spatial architectures are preferred when local features and relationships are most important, while spectral methods excel in capturing global graph structures. Some recent models combine both approaches to leverage the strengths of each, leading to hybrid models that offer both local and global feature learning capabilities.

1.4 Advanced GNN Architectures

GNNs have experienced rapid growth, resulting in a variety of advanced architectures that extend the fundamental concepts of graph-based learning to address increasingly complex and diverse types of graph data. These advanced architectures not only improve the expressiveness and scalability of GNNs but also enable them to handle specific challenges, such as multi-relational data, large graphs, temporal changes, and graph isomorphisms. This section focuses on three major advancements in GNNs: Graph Isomorphism Networks (GIN), Relational Graph Convolutional Networks (R-GCN), and GraphSAGE: Inductive Representation Learning. Additionally, we will also touch upon Dynamic and Temporal GNNs, which aim to model graphs that change over time.

These architectures play a crucial role in overcoming limitations faced by traditional GNNs and are essential for applications such as graph classification, link prediction, and knowledge graph embedding. Below, we discuss each architecture in detail, exploring their underlying principles, mathematical formulations, and key advantages.

1.4.1 Graph Isomorphism Networks (GIN)

In order to overcome the shortcomings of conventional GNNs in identifying non-isomorphic graphs, GINs were developed. In traditional GNNs, the aggregation functions commonly used to combine information from neighboring nodes, such as summing or averaging node features, often lack the expressive power required to differentiate graphs that may have different structures but are not easily distinguished by these basic operations. This limitation becomes apparent when dealing with graph classification tasks where the model needs to distinguish between graphs that are structurally different but share similar neighborhood structures.

GINs get over this restriction by implementing a stronger aggregation technique. Utilizing a sum aggregation function, which aggregates the features of nearby nodes by adding them together, is the fundamental concept underlying GIN. A multi-layer perceptron (MLP) is then used to interpret the aggregated data. By using this method, the network is guaranteed to be able to identify tiny variations in graph architecture and capture more intricate interactions between nodes and their neighbors [39]. The sum aggregation function allows GIN to fully encode structural information, while the MLP refines this aggregated information, increasing the model's expressiveness. Mathematically, the aggregation process in GIN is expressed as Eq. (1.5):

$$\mathrm{h}_v^{(k)} = MLP\left(\mathrm{Aggregate}(\mathrm{h}_u^{(k-1)} | u \in \mathrm{N}(v))\right), \tag{1.5}$$

Here, $h_v^{(k)}$ is the feature of node v at the k-th layer, and $N(v)$ denotes the set of neighbors of node v. The aggregation function sums the features of node v 's neighbors, and the resulting sum is passed through an *MLP* to obtain the updated feature for node v. One of the key strengths of GIN is its expressiveness. The sum aggregation method allows the model to capture structural information more effectively, making it capable of distinguishing between non-isomorphic graphs. In fact, GIN has been proven to be as powerful as the Weisfeiler-Lehman graph isomorphism test, which is a classical method for testing graph isomorphism. This makes GIN particularly useful for tasks that require the ability to discern subtle differences in graph structure. Additionally, GIN models are scalable and can be applied to large graphs, as the primary computation involves simple summation operations over neighboring nodes [40].

GINs have been particularly effective in applications such as chemoinformatics, where molecular structures are represented as graphs and need to be classified, and in social network analysis, where the goal is to understand and classify different social structures. Other applications include protein-protein interaction networks, where distinguishing between various interaction patterns is critical. By improving the ability to differentiate graph structures, GINs significantly enhance the performance of GNNs in complex graph-based tasks.

1.4.2 Relational Graph Convolutional Networks (R-GCN)

R-GCNs are an extension of traditional GCNs, specifically designed to handle multi-relational graph data where edges between nodes can have different types or roles. In many real-world scenarios, such as knowledge graphs or social networks, the relationships between entities are not uniform. For example, in a knowledge graph, entities like "person" and "organization" could be linked by multiple types of relationships, such as "works_at", "affiliated_with", or "owns". Traditional GCNs struggle to capture the diversity of such relationships, as they treat all edges uniformly. R-GCNs address this issue by introducing a separate convolution for each edge type, allowing them to capture the unique characteristics and semantics of each relationship.

The core idea behind R-GCNs is to learn different transformations for each type of edge in the graph. Instead of performing a single graph convolution, as in traditional GCNs, R-GCNs use edge-specific weight matrices to aggregate information from neighbors connected through different types of edges [41]. The node feature update rule in an R-GCN is given by Eq. (1.6):

$$h_v^{(k)} = \sigma\left(\sum_{r \in R}\sum_{u \in N_r(v)} \frac{1}{c_r} W_r^{(k)} h_u^{(k-1)}\right), \tag{1.6}$$

Here, $h_v^{(k)}$ represents the feature of node v at layer k, and the summation is performed over all edge types r and their corresponding neighbors u. $W_r^{(k)}$ is the weight matrix for edge type r at layer k, and c_r is a normalization constant to ensure that the contributions from different edge types are balanced. By using separate weight matrices for each edge type, R-GCNs can learn how different relationships affect the node features in a distinct manner. R-GCNs are particularly useful for tasks that involve graphs with multiple types of relationships. For example, in knowledge graph completion, R-GCNs can predict missing relationships between entities based on the existing graph structure. Similarly, in recommendation systems, R-GCNs can model the different types of interactions between users and items (such as "viewed", "purchased", or "rated") to make personalized predictions [42]. These capabilities make R-GCNs highly effective for tasks that require modeling complex and heterogeneous graphs.

One of the main challenges of R-GCNs is efficiently scaling to large graphs with numerous edge types. As the number of edge types increases, the number of parameters required to learn separate transformations for each edge type also grows. Despite this, R-GCNs have demonstrated strong performance in handling multi-relational data and have become an important tool in areas such as knowledge graph completion, social network analysis, and recommender systems.

1.4.3 GraphSAGE: Inductive Representation Learning

GraphSAGE (Graph Sample and Aggregation) is an influential method in the field of GNNs that tackles the problem of inductive representation learning on large, unseen graphs. Unlike traditional GNNs, which typically require the entire graph to be available during training, GraphSAGE enables inductive learning, meaning it can generalize to unseen nodes and edges after training on a subgraph. This ability to learn from large graphs without needing to process the entire graph at once is crucial for applications in which the graph evolves dynamically or the graph is too large to fit into memory. The core idea of GraphSAGE is to aggregate information from a node's local neighborhood to compute its representation. Instead of using the entire graph structure during training, GraphSAGE samples a fixed-size neighborhood around each node and performs aggregation over this neighborhood to generate the node's embedding. This aggregation step ensures that the model can capture the structure and relationships within a node's local vicinity without needing access to the full graph [43]

The architecture of GraphSAGE consists of multiple layers, where each layer aggregates information from a node's neighbors and combines it with its own feature. The standard GraphSAGE framework can be mathematically described by the following update rule as Eq. (1.7):

$$h_v^{(k)} = \sigma\left(W^{(k)}.\text{AGGREGATE}^{(k)}\left(\left\{h_u^{(k-1)} : u \in N(v)\right\}\right)\right), \tag{1.7}$$

where:

- AGGREGATE$^{(k)}$ is the aggregation function that combines the features of node v with the features of its neighbors. This function can vary and is typically chosen to be one of several options, such as mean, sum, or pooling.
- $W^{(k)}$ is a weight matrix for the k-th layer.

The aggregation function is crucial in GraphSAGE, as it determines how the information from the neighbors is combined. Several aggregation methods have been proposed, including:

(i) Mean Aggregation: A simple method where the feature vectors of the neighbors are averaged as described in Eq. (1.8).

$$\text{AGGREGATE}^{(k)} = \frac{1}{|N(v)|} \sum_{u \in N(v)} h_u^{(k-1)}, \tag{1.8}$$

(ii) LSTM-based Aggregation: Uses a Long Short-Term Memory (LSTM) network to aggregate neighbors' information, enabling the model to learn more complex relationships in the neighborhood.
(iii) Pooling Aggregation: Combines neighbor features using a pooling operation, such as max-pooling.

GraphSAGE employs these aggregation methods to handle various graph structures and improve the quality of node embeddings for inductive learning tasks. It has proven effective in a variety of applications. One prominent example is in social network analysis, where new users continuously join the platform, and the graph of users and their relationships evolves over time. GraphSAGE allows for efficient learning and representation of new nodes (users) without retraining on the entire graph. It also excels in scenarios such as recommendation systems, where it can generate embeddings for previously unseen items or users by leveraging their local graph context [44].

The ability to perform inductive learning makes GraphSAGE especially useful for large-scale graphs and dynamic systems where the graph structure is continuously changing. This capacity to generalize to new, unseen data without the need for retraining on the entire graph is one of GraphSAGE's defining features, distinguishing it from traditional, transductive GNNs.

1.5 Algorithms for GNNs

The capacity of GNNs to simulate intricate graph-structured data, including social networks, chemical structures, and recommendation systems, has drawn a lot of interest. Effective algorithms are needed to manage the difficulties presented by large-scale graphs, a variety of graph architectures, and intricate relationships in order to fully utilize GNNs in a range of real-world applications. The main methods and algorithms utilized in GNNs to enhance their scalability and performance will be covered in this section.

We will start by exploring aggregation and update mechanisms, which are essential for capturing neighborhood information and updating node features. Then, we will discuss optimization techniques that ensure the effective training of GNNs, especially on large graphs. Finally, we will dive into sampling methods that allow GNNs to handle large-scale graphs by reducing computational complexity. These methods include neighborhood sampling and graph partitioning approaches that enhance the scalability of GNNs.

1.5.1 Aggregation and Update Mechanisms

Aggregation and update mechanisms are at the core of GNNs and are essential for enabling the model to capture the structural information inherent in graph-structured data. In GNNs, each node aggregates information from its neighbors and updates its feature representation based on this aggregated information. This process allows the model to propagate information throughout the graph, facilitating the learning of meaningful node, edge, and graph-level representations. The aggregation step typically involves combining the feature vectors of a node's neighbors. The most used aggregation functions include summation, mean, and max pooling. Each function has its strengths and can be chosen depending on the problem [45]. For example, the sum function aggregates the features by adding them together, which can help preserve the scale of the feature values. In contrast, the mean function normalizes the aggregated feature, providing a balanced representation. Max pooling aggregates by taking the maximum value for each feature dimension, focusing on the most prominent features.

Once the information from neighboring nodes has been aggregated, the next step is to update the node's representation. This is typically done by applying a neural network layer, such as a fully connected layer (also known as a perceptron), to the aggregated features. The update function typically includes a non-linear activation function (e.g., ReLU) to introduce non-linearity and enable the model to learn more complex patterns [46]. Mathematically, the aggregation and update steps can be formalized as Eqs. (1.9) and (1.10):

(i) Aggregation:

For each node v_i in the graph, the feature vector $\mathrm{h}_i^{(k)}$ at the k-th layer is updated by aggregating information from its neighbors $\mathrm{N}(v_i)$:

$$\mathrm{h}_i^{\sim(k)} = \mathrm{AGGREGATE}\left(\left\{\mathrm{h}_j^{(k-1)} : j \in \mathrm{N}\left(v_i\right)\right\}\right), \tag{1.9}$$

where $\tilde{h}_i^{(k)}$ is the aggregated feature vector of node v_i at the k-th layer, and $N(v_i)$ represents the neighbors of node v_i.

(ii) Update:

After aggregation, the node's feature is updated as:

$$h_i^{(k)} = \sigma\left(W^{(k)}\tilde{h}_i^{(k)} + b^{(k)}\right), \tag{1.10}$$

where $W^{(k)}$ is the weight matrix, $b^{(k)}$ is the bias term, and σ is the activation function (such as ReLU).

The aggregation and update procedures allow GNNs to spread information across the graph and are essential for learning node embeddings. GNNs are useful for node classification, link prediction, and graph classification because these methods enable the network to recognize both local and global structural patterns. GNNs are more adaptable and able to handle a variety of graph shapes in practice because of more complex aggregation approaches, such as attention mechanisms in GATs, which enable weighted aggregation based on the relevance of each neighbor. Additionally, innovations like graph pooling and readout functions help summarize node representations at higher levels, enabling the model to learn graph-level representations for tasks like graph classification.

1.5.2 Optimization Techniques for GNN Training

Training GNNs involves optimizing the model's parameters to accurately learn the graph's structure and features. GNNs face unique challenges due to their dependence on graph-structured data, which can be large and sparse. Optimizing GNNs requires specialized techniques that address both the challenges of graph-based data and the computational complexity involved in training deep models. Several optimization strategies have been developed to efficiently train GNNs, improving convergence, scalability, and generalization.

One of the key challenges in GNN optimization is the localization of computation. Unlike traditional neural networks, which process data in fixed dimensions (e.g., images or text), GNNs operate on graph-structured data, where the topology of the graph can vary and change over time. Mini-batch training is commonly employed to optimize training, where the graph is partitioned into smaller, and only a subset of nodes and edges is used for each update. This helps reduce the memory requirements and computational load, making it feasible to train GNNs on large graphs [47]. A widely used method for optimizing GNNs is stochastic gradient descent (SGD) and its variants, including Adam and RMSprop, which adapt learning rates during training. These optimizers adjust the learning rate dynamically, which helps prevent oscillations in the optimization process and enables faster convergence [48]. Moreover, weight decay or L2 regularization is often applied to control overfitting and prevent the model from becoming overly complex by penalizing large weights. This technique helps generalize the model and reduces the risk of overfitting to the training data. Another optimization strategy in GNNs is gradient clipping, which is used to mitigate the issue of exploding gradients in deep networks. Since GNNs involve recursive message passing between nodes, deep GNNs are prone to large gradients during backpropagation. Gradient clipping involves limiting the magnitude of the gradients to a predefined threshold, preventing large updates that can destabilize the training process [49].

Additionally, sampling techniques such as neighbor and subgraph sampling have been proposed in the context of large-scale graphs. These methods reduce the amount of data that needs to be processed in each training step. For example, in the GraphSAGE model, the neighbors of each node are sampled to create mini-batches, allowing for scalable training on large graphs. These techniques are essential for enabling the training of GNNs on massive datasets, such as social networks or citation graphs, where the graph size can be enormous [50]. Mathematically, the optimization process typically involves the minimization of a loss function L over the graph, where the objective is to find the model parameters θ that minimize the loss concerning the training data as Eqs. (1.11) and (1.12):

$$\theta^* = \arg\min_{\theta} \sum_i L\left(h_i, y_i\right), \tag{1.11}$$

where h_i is the learned representation of the node i, y_i is the target label, and $L(h_i, y_i)$ is the loss function. The optimization is performed using an algorithm like Adam, which updates the model parameters as follows:

$$\theta^{(t+1)} = \theta^{(t)} - \eta . \hat{g}^{(t)}, \tag{1.12}$$

where η is the learning rate and $\hat{g}^{(t)}$ is the gradient of the loss function at the time step t, which is computed using backpropagation.

In addition to these techniques, other advanced optimization methods, such as early stopping, learning rate decay, and adaptive optimization strategies, are commonly used in practice to improve the efficiency of training and enhance model performance. The optimization of GNNs involves addressing both the computational and structural challenges posed by graph data. By employing strategies such as mini-batch training, stochastic gradient-based optimization, and sampling techniques, GNNs can be trained efficiently on large-scale graphs. These techniques, combined with regularization methods, help enhance the scalability and generalization of GNN models, making them more effective in real-world applications.

1.5.3 Sampling Methods for Large Graphs

When working with large-scale graphs, directly applying GNNs to the entire graph can be computationally expensive and infeasible due to memory and processing limitations. A key challenge in training GNNs on large graphs is the issue of scalability. Processing the entire graph at once may require a prohibitively large amount of memory and time. To address this challenge, sampling methods have been introduced to reduce the size of the graph that needs to be processed at each step of training. Sampling methods help in creating mini-batches of graph data, making it possible to train on large graphs efficiently.

Sampling techniques aim to extract a subset of nodes, edges, or subgraphs that capture important structural and feature-based information while reducing the computational burden. These methods are essential for inductive learning, where the goal is to make predictions on unseen data or graphs with unseen nodes.

1.5.3.1 Neighbor Sampling

Neighbor sampling is one of the most widely used techniques for GNNs, particularly for large graphs. In this method, for each node, a fixed number of neighboring nodes are sampled to create a mini-batch. This reduces the number of nodes and edges that need to be processed, which helps to mitigate the high memory and computational costs of processing the entire graph.

A common implementation of neighbor sampling is used in the GraphSAGE model. In GraphSAGE, instead of using the entire neighborhood of a node for message passing, a fixed-size neighborhood is sampled [51]. For each node v, a set of k neighbors is randomly selected, and their features are aggregated to update v 's representation. This sampling process reduces the complexity from $O(|V|^2)$ to $O(k|V|)$, where k is the number of neighbors sampled, and $|V|$ is the number of nodes in the graph.

Mathematically, the node representation update in neighbor sampling can be represented as Eq. (1.13):

$$\mathrm{h}_v^{(l+1)} = \mathrm{Aggregate}\left(\left\{\mathrm{h}_u^{(l)} : u \in \mathrm{N}(v)\right\}\right), \tag{1.13}$$

where $\mathrm{h}_v^{(l+1)}$ is the updated representation of the node v at layer $l + 1$, $\mathrm{N}(v)$ is the set of neighbors of the node v, and the aggregate function combines the features of the sampled neighbors.

1.5.3.2 Subgraph Sampling

Subgraph sampling involves extracting subgraphs (connected components or neighborhood patches) from the original graph to create smaller, more manageable chunks. Subgraph sampling works by selecting a set of nodes and including all the nodes that are connected to them within a specific neighborhood radius. By sampling subgraphs, the GNN can learn localized structures while avoiding the need to process the entire graph.

This method is particularly useful when dealing with graph-wide dependencies (such as in large social networks or citation graphs), as it allows the model to focus on local neighborhood structures rather than the full graph. GraphSAGE and other models like FastGCN use subgraph sampling to enable inductive learning on graphs of unseen size. In subgraph sampling, the training process involves extracting a subgraph around each node v, including v 's neighbors up to a certain hop distance, forming a local subgraph $G_{v,d}$ of depth d [52]. The GNN model is then trained on these subgraphs instead of the entire graph. This process is repeated for each node in the graph.

1.5.3.3 Random Walks and Metropolis-Hastings Sampling

Another important sampling method is random walks, where the training process selects paths or sequences of nodes by randomly walking through the graph. Starting at a specific node, a random walk iteratively chooses nearby nodes based on predetermined probabilities. Long-range dependencies between distant nodes in a graph can be captured with this method.

In addition to simple random walks, Metropolis-Hasting's sampling can be applied to sample nodes more effectively, particularly in graphs with biased or skewed distributions. In this case, the random walk is modified based on a certain acceptance probability to ensure that rare nodes or edges are included in the sample [53]. These methods allow for a more controlled exploration of the graph and are especially beneficial for graphs where not all nodes or edges need to be included in every training step.

1.5.3.4 GraphSAGE Sampling

GraphSAGE introduces a neighborhood sampling scheme to handle inductive learning. Instead of using the entire graph for each training example, GraphSAGE samples a fixed number of neighbors from each node during training. The sampled neighbors are then aggregated to form the updated node representation [54]. This sampling process allows GraphSAGE to scale effectively to large graphs, as the computation complexity per node becomes independent of the graph size.

The sampling process in GraphSAGE is as follows:

(i) For each node v, sample a fixed number of neighbors from its local neighborhood.
(ii) Aggregate the features of the sampled neighbors using an aggregation function, such as mean, sum, or pooling.
(iii) Update the node's feature representation based on the aggregated information.

1.5.3.5 Importance Sampling

Importance sampling is another technique where nodes or edges are sampled based on their importance in the graph. The importance can be computed according to certain graph metrics, such as degree centrality or closeness centrality, which identify influential nodes in the network [55]. By prioritizing sampling nodes that are more critical to the overall graph structure, importance sampling allows the model to focus on more relevant parts of the graph, thus improving learning efficiency.

1.5.3.6 Node Sampling

In node sampling, instead of sampling neighbors, a set of nodes is selected randomly or based on specific criteria. Each node in the sample contributes to the model's learning process. This method is often used in combination with other sampling techniques, such as neighbor sampling or subgraph sampling, to improve computational efficiency. Sampling methods for GNNs offer a variety of solutions for dealing with large-scale graph data. Neighbor sampling, subgraph sampling, random walks, and importance sampling all play vital roles in reducing the computational complexity of GNNs while maintaining model accuracy and performance [56]. By applying these techniques, GNNs can be trained efficiently on large graphs, enabling them to handle real-world applications such as social networks, knowledge graphs, and large-scale recommendation systems.

1.6 Challenges and Future Directions

GNNs have shown significant promise in various fields, but they still face several challenges that hinder their full potential. One of the major challenges is scalability and efficiency. As GNNs become deeper and handle more complex graph structures, their computational complexity increases, especially for large-scale graphs with millions or even billions of nodes and edges. Despite efforts like sampling methods, improving the scalability of GNNs remains a key issue for their practical application. Another challenge is the problem of over-smoothing in deep GNNs. When too many layers aggregate information from neighbors, the node representations can become indistinguishable, losing valuable structural information. Addressing over-smoothing while maintaining effective learning at deeper levels of the network is a critical area for improvement.

Additionally, GNNs struggle with generalization, particularly in inductive learning tasks where the model must make predictions on graphs or subgraphs it has never seen before. This issue is compounded by the sensitivity of GNNs to noisy or incomplete data, which can affect their robustness in real-world applications. Another open problem lies in handling heterogeneous graphs with multiple types of nodes and edges. Real-world graphs often have complex relationships, and current methods may not be equipped to handle the variety of connections found in domains like knowledge graphs or social networks. Furthermore, GNNs are often treated as black-box models, which leads to a lack of interpretability and explainability. In sensitive domains such as healthcare, where understanding the reasoning behind predictions is essential, enhancing the transparency and interpretability of GNNs is crucial. Looking forward, there are several exciting directions for future research. One area of focus will be improving the scalability of GNNs without compromising their performance. Innovations in efficient sampling techniques, hierarchical graph representations, and distributed learning could enable GNNs to handle large-scale datasets more effectively. Incorporating temporal or dynamic data into GNNs will also be a critical area of development. Many real-world graphs evolve, and GNNs must be able to adapt to these changes, especially in domains like social networks, transportation systems, and financial markets. Additionally, the rise of self-supervised and contrastive learning methods in GNNs presents a promising future. These techniques allow models to learn useful representations from unlabeled data, which can be particularly useful in scenarios where labeled data is scarce.

The integration of GNNs with other deep learning models, such as CNNs or reinforcement learning, is another exciting avenue for future research. Hybrid models that combine the strengths of multiple architectures could result in more powerful systems capable of solving complex, multi-task problems. Finally, improving the explainability and fairness of GNNs will be an essential focus, especially as these models are deployed in critical areas. Ensuring transparency in decision-making and reducing bias will be necessary for the responsible deployment of GNNs in real-world applications. In summary, while GNNs have shown great promise, there are still significant challenges that need to be addressed. By advancing the scalability, generalization, and interpretability of GNNs, as well as exploring new research directions such as dynamic graphs and hybrid models, GNNs have the potential to revolutionize a variety of domains, from healthcare to finance to social network analysis.

1.7 Conclusion

In this chapter, we have explored the fundamental concepts, architectures, and advanced techniques of GNNs, highlighting their transformative impact on machine learning and their ability to handle graph-structured data. GNNs have revolutionized the way we approach problems where relationships and dependencies between data points are paramount. The message-passing mechanism and node aggregation processes inherent in GNNs enable these networks to capture both local and global dependencies within graphs, making them highly effective in domains such as social network analysis, recommendation systems, drug discovery, and more. Through the in-depth analysis of various GNN architectures, including GCNs, GATs, and GraphSAGE, we have illustrated their unique strengths, applications, and the trade-offs that come with different modeling strategies.

As the field continues to evolve, several challenges and open problems remain. One significant challenge is the scalability of GNNs for large graphs. While methods like sampling and optimization strategies have been proposed to address this, efficient processing of large-scale graphs remains an ongoing research problem.

Another challenge is the generalization of GNNs across different types of graph structures, particularly in dynamic or evolving graphs. Additionally, there is still debate regarding the interpretability of GNN models, despite the fact that real-world applications depend on our ability to comprehend how these models make decisions, particularly in complicated graphs. There are intriguing prospects to get around some of these issues by combining GNNs with other deep learning models, such as reinforcement learning and attention processes. The future of GNNs is bright, with substantial opportunities for further advancements in both research and industry. As GNN models continue to improve, there will be a focus on developing more efficient algorithms, better handling of heterogeneous graphs, and improved inductive learning techniques that allow GNNs to perform well on unseen data. Additionally, the incorporation of self-supervised learning methods and contrastive learning could enhance the ability of GNNs to leverage unlabeled data, opening new possibilities in semi-supervised and unsupervised learning.

In industry, the applications of GNNs will continue to expand across domains such as healthcare, finance, transportation, and beyond. Their ability to model complex relationships in data and generate insights from graph-structured data will drive innovation in areas like personalized medicine, fraud detection, and autonomous systems. Furthermore, the potential of GNNs in multi-modal and multi-task learning, as well as their application in emerging fields such as quantum computing, will likely redefine the boundaries of what is possible in artificial intelligence. Overall, the growth of GNNs holds transformative potential, both in advancing theoretical research and in practical applications. With ongoing research addressing current limitations, the future of GNNs is promising, with their ability to revolutionize industries and create new opportunities for innovation across diverse sectors. As we continue to push the boundaries of what GNNs can achieve, their ability to capture intricate, real-world relationships in data will undoubtedly remain one of the most powerful tools in the machine learning toolbox.

References

1. Ga S, et al. Efficient GNN-based social recommender systems through social graph refinement. J Supercomput. 2024;81, 2024(1) https://doi.org/10.1007/s11227-024-06682-w.
2. Cao Y, et al. A data-centric framework of improving graph neural networks for knowledge graph embedding. World Wide Web. 2024;28(1) https://doi.org/10.1007/s11280-024-01320-0.
3. Liu J, Chen S, Shen L, Liu J, Chen S, Shen L. A comprehensive survey on graph neural network accelerators. Front Comput Sci. 2024;2025, 19(2) https://doi.org/10.1007/s11704-023-3307-2.
4. Zhuang W, et al. Approximate message passing-enhanced graph neural network for OTFS data detection, IEEE Journals & Magazine | IEEE Xplore. IEEE Wireless Commun Lett. 2024;13(7) https://doi.org/10.1109/LWC.2024.3395700.
5. Li Z, Ma J. Learning feature matching via Matchable Keypoint-assisted graph neural network | IEEE Journals & Magazine | IEEE Xplore. IEEE Trans Image Process. 2025;34 https://doi.org/10.1109/TIP.2024.3512352.
6. Liu Z, et al. Proformer: a scalable graph transformer with linear complexity. Appl Intell. 2024;55, 2024(2) https://doi.org/10.1007/s10489-024-06065-x.

7. Yingxia S, et al. Distributed graph neural network training: a survey. ACM Comput Surv. 2024;56(8) https://doi.org/10.1145/3648358.
8. Kartik S, et al. A survey of graph neural networks for social recommender systems. ACM Comput Surv. 2024;56(10) https://doi.org/10.1145/3661821.
9. Hasan M, et al. Enhancing stroke disease classification through machine learning models via a novel voting system by feature selection techniques. PLoS One. 2025;20(1):e0312914. https://doi.org/10.1371/journal.pone.0312914.
10. Lai Y, Zhu Y, Fan W, Zhang X, Zhou K. Toward Adversarially robust recommendation from adaptive fraudster detection | IEEE Journals & Magazine | IEEE Xplore. IEEE Trans Inf Forensics Secur. 2024;19 https://doi.org/10.1109/TIFS.2023.3327876.
11. Esmaeili B, Azmoodeh A, Dehghantanha A, Srivastava G, Karimipour H, Lin JC-W. A GNN-based adversarial internet of things malware detection framework for critical infrastructure: studying Gafgyt, Mirai, and tsunami campaigns | IEEE Journals & Magazine | IEEE Xplore. IEEE Internet Things J. 2024;11(16) https://doi.org/10.1109/JIOT.2023.3298663.
12. Jiang W, et al. Graph neural networks for routing optimization: challenges and opportunities. Sustainability. 2024;16(21):9239. https://doi.org/10.3390/su16219239.
13. Wang R, et al. Adaptive multi-channel deep graph neural networks. Symmetry. 2024;16(4):406. https://doi.org/10.3390/sym16040406.
14. Peng X, Wei Z, Li C, Yuan J, Liu Z, Liu W. Drug-target prediction based on dynamic heterogeneous graph convolutional network | IEEE Journals & Magazine | IEEE Xplore. IEEE J Biomed Health Inform. 2024;28(11) https://doi.org/10.1109/JBHI.2024.3441324.
15. Arshad Choudhry I, et al. A novel interpretable graph convolutional neural network for multimodal brain tumor segmentation. Cogn Comput. 2024;17(1) https://doi.org/10.1007/s12559-024-10387-w.
16. Yan Y, et al. A review of graph theory-based diagnosis of neurological disorders based on EEG and MRI. Neurocomputing. 2024;599 https://doi.org/10.1016/j.neucom.2024.128098.
17. Gang L, Eric I, Tengfei L, Meng J. Rationalizing graph neural networks with data augmentation. ACM Trans Know Discover Data. Feb. 2024;18(4) https://doi.org/10.1145/3638781.
18. He H, Xianghao Y, Zhang J, Song S, Letaief KB. Message passing meets graph neural networks: a new paradigm for massive MIMO systems | IEEE Journals & Magazine | IEEE Xplore. IEEE Trans Wirel Commun. 2024;23(5) https://doi.org/10.1109/TWC.2023.3321667.
19. Wang Y, Liang H, Cao X, Chang Y, Tsang IW. Enhancing locally adaptive smoothing of graph neural networks via Laplacian node disagreement | IEEE Journals & Magazine | IEEE Xplore. IEEE Trans Knowl Data Eng. 2024;36(3) https://doi.org/10.1109/TKDE.2023.3303212.
20. Nazzal M, Khreishah A, Lee J, Angizi S, Al-Fuqaha A, Guizani M. Semi-decentralized inference in heterogeneous graph neural networks for traffic demand forecasting: an edge-computing approach | IEEE Journals & Magazine | IEEE Xplore. IEEE Trans Veh Technol. 2024;73(12) https://doi.org/10.1109/TVT.2024.3355971.
21. Shetty RD, Bhattacharjee S, Thanmai K. Node classification in weighted complex networks using neighborhood feature similarity | IEEE Journals & Magazine | IEEE Xplore. IEEE Trans Emerg Topics Comput Intell. 2024;8(6) https://doi.org/10.1109/TETCI.2024.3380481.
22. Yasmin F, et al. PoxNet22: a fine-tuned model for the classification of Monkeypox disease using transfer learning. IEEE Access. 2023;11:24053–76. https://doi.org/10.1109/ACCESS.2023.3253868.
23. Yasmin F, et al. AQIPred: a hybrid model for high precision time specific forecasting of air quality index with cluster analysis. Hum-Cent Intell Syst. 2023;3(3):275–95. https://doi.org/10.1007/s44230-023-00039-x.
24. Guo J, Huang K, Zhang R, Yi X. ES-GNN: generalizing graph neural networks beyond Homophily with edge splitting | IEEE Journals & Magazine | IEEE Xplore. IEEE Trans Pattern Anal Mach Intell. 2024;46(12) https://doi.org/10.1109/TPAMI.2024.3459932.
25. Yu R, Li Y, Liang H, Chen Z. GeoExplainer: interpreting graph convolutional networks with geometric masking. Neurocomputing. 2024;605 https://doi.org/10.1016/j.neucom.2024.128393.

26. Wang J, Guo Y, Yang L, Wang Y. Binary graph convolutional network with capacity exploration. IEEE Trans Pattern Anal Mach Intell. 2024;46(5):3031–46. https://doi.org/10.1109/TPAMI.2023.3342224.
27. Lv S, Cheng Q, Gong H, Gao H, Zhou D, Zheng D. Developing a silicon beam emulator using graph attention networks. IEEE Access. 2024;12:179119–29. https://doi.org/10.1109/ACCESS.2024.3505603.
28. Ben J, Sun Q, Liu K, Yang X, Zhang F. Multi-head multi-order graph attention networks. Appl Intell. 2024;54(17–18):8092–107. https://doi.org/10.1007/s10489-024-05601-z.
29. Chen T, Li H, Zheng H, Chen J, Fan Y. dFCExpert: learning dynamic functional connectivity patterns with modularity and state experts. bioRxiv. 2024; https://doi.org/10.1101/2024.12.20.629773.
30. Yuan J, Chen S, Mo B, Ma Y, Zheng W, Zhang C. R-GNN: recurrent graph neural networks for font classification of oracle bone inscriptions. Herit Sci. 2024;12(1):30. https://doi.org/10.1186/s40494-024-01133-4.
31. Yasmin F, Hassan MM, Zaman S, Aung ST, Karim A, Azam S. A forecasting prognosis of the Monkeypox outbreak based on a comprehensive statistical and regression analysis. Computation. 2022;10(10):10. https://doi.org/10.3390/computation10100177.
32. Zhang W, et al. A novel soft actor-critic framework with disjunctive graph embedding and autoencoder mechanism for job shop scheduling problems. J Manuf Syst. 2024;76:614–26. https://doi.org/10.1016/j.jmsy.2024.08.015.
33. Zhu R, Peng J, Huang W, He Y, Tang C. Dynamic graph attention-guided graph clustering with entropy minimization self-supervision. Appl Intell. 2024;54(24):12819–34. https://doi.org/10.1007/s10489-024-05745-y.
34. Zheng K, Yu S, Chen B. CI-GNN: a granger causality-inspired graph neural network for interpretable brain network-based psychiatric diagnosis. Neural Netw. 2024;172:106147. https://doi.org/10.1016/j.neunet.2024.106147.
35. Liu Z, Chen Q, Lan W, Lu H, Zhang S. SSLDTI: a novel method for drug-target interaction prediction based on self-supervised learning. Artif Intell Med. 2024;149:102778. https://doi.org/10.1016/j.artmed.2024.102778.
36. Tian H, Zhang X, Zheng X, Zhang Z, Zeng DD. Graph representation learning of multilayer spatial-temporal networks for stock predictions. IEEE Trans Comput Soc Sys. 2024; https://doi.org/10.1109/TCSS.2024.3459945.
37. Mohammadi S, Allali M. Advancing brain tumor segmentation with spectral-spatial graph neural networks. Appl Sci-Basel. 2024;14(8):3424. https://doi.org/10.3390/app14083424.
38. Zha Y, Yang Y. Innovative graph neural network approach for predicting soil heavy metal pollution in the Pearl River Basin, China. Sci Rep. 2024;14(1):16505. https://doi.org/10.1038/s41598-024-67175-7.
39. Li Q, Yu JX. Fast local subgraph counting. Proc VLDB Endow. 2024;17(8):1967–80. https://doi.org/10.14778/3659437.3659451.
40. Wu C, Lou Y, Li J, Wang L, Xie S, Chen G. A multitask network robustness analysis system based on the graph isomorphism network. IEEE T Cybern. 2024;54(11):6630–42. https://doi.org/10.1109/TCYB.2024.3422430.
41. Khemani B, Patil S, Kotecha K, Tanwar S. A review of graph neural networks: concepts, architectures, techniques, challenges, datasets, applications, and future directions. J Big Data. 2024;11(1):18. https://doi.org/10.1186/s40537-023-00876-4.
42. Li S, Zhang R. A novel interactive deep cascade spectral graph convolutional network with multi-relational graphs for disease prediction. Neural Netw. 2024;175:106285. https://doi.org/10.1016/j.neunet.2024.106285.
43. Jiawei E, Zhang Y, Yang S, Wang H, Xia X, Xu X. GraphSAGE plus plus: weighted multiscale GNN for graph representation learning. Neural Process Lett. 2024;56(1):24. https://doi.org/10.1007/s11063-024-11496-1.

44. Ning D, Vetrova V, Bryan KR, Koh YS. Harnessing the power of graph representation in climate forecasting: predicting global monthly mean sea surface temperatures and anomalies. Earth Space Sci. 2024;11(3):e2023EA003455. https://doi.org/10.1029/2023EA003455.
45. Xu G, Rao G, Zhang L, Cong Q. Entity-relation aggregation mechanism graph neural network for knowledge graph embedding. Appl Intell. 2025;55(1):43. https://doi.org/10.1007/s10489-024-05907-y.
46. Liu Y, Li H, Hao M. Personalized and privacy-preserving federated graph neural network. Front Phys. 2024;12:1383276. https://doi.org/10.3389/fphy.2024.1383276.
47. Wang G, Warrell J, Zheng S, Gerstein M. A variational graph-partitioning approach to modeling protein liquid-liquid phase separation. Cell Rep Phys Sci. 2024;5(11):102292. https://doi.org/10.1016/j.xcrp.2024.102292.
48. Yasmin F, et al. GastroNet: gastrointestinal polyp and abnormal feature detection and classification with deep learning approach. IEEE Access. 2023;11:97605–24. https://doi.org/10.1109/ACCESS.2023.3312729.
49. Li Y, Song X, Gong K, Liu S, Li W. Differentially private graph neural networks for graph classification and its adaptive optimization. Expert Syst Appl. 2025;263:125798. https://doi.org/10.1016/j.eswa.2024.125798.
50. Zhang S, Hu Z, Subramonian A, Sun Y. Motif-driven contrastive learning of graph representations. IEEE Trans Knowl Data Eng. 2024;36(8):4063–75. https://doi.org/10.1109/TKDE.2024.3364059.
51. Song Y, Chen PH, Lu Y, Abrar N, Kalavri V. In situ neighborhood sampling for large-scale GNN training. In: *Proceedings of the 20th international workshop on data management on new hardware*, in DaMoN '24. New York: Association for Computing Machinery; 2024. p. 1–5. https://doi.org/10.1145/3662010.3663443.
52. Ye Y, Lian X, Chen M. Efficient exact subgraph matching via GNN-based path dominance embedding. Proc VLDB Endow. 2024;17(7):1628–41. https://doi.org/10.14778/3654621.3654630.
53. Park J, Kang N. BMO-GNN: Bayesian mesh optimization for graph neural networks to enhance engineering performance prediction. J Comput Design Eng. 2024;11(6):260–71. https://doi.org/10.1093/jcde/qwae102.
54. Huang K, Chen C. Subgraph generation applied in GraphSAGE deal with imbalanced node classification. Soft Comput. 2024;28(17):10727–40. https://doi.org/10.1007/s00500-024-09797-7.
55. Sun K, Jiang H, Hu Y, Yin B. Incorporating Multi-Level Sampling with Adaptive Aggregation for Inductive Knowledge Graph Completion. ACM Trans Knowl Discov Data. 2024;18(5):128:1–128:16. https://doi.org/10.1145/3644822.
56. SCN_GNN: A GNN-based fraud detection algorithm combining strong node and graph topology information. Expert Syst Appl. 2024;237:121643. https://doi.org/10.1016/j.eswa.2023.121643.

Chapter 2
Fundamentals of Machine Learning in Neurology

Md Rittique Alam and Abdullah Al Abid

2.1 Introduction

Neurological disorders pose a considerable and escalating global health challenge, impacting more than 3 billion individuals around the world as of 2021. These conditions include a range of diseases that affect the brain, spinal cord, and nerves across the body, resulting in varying levels of disability and impairment. The World Health Organization (WHO) emphasizes that neurological disorders have become the primary cause of ill health and disability, exceeding other health concerns in both prevalence and their effect on quality of life [1, 2]. Common neurological disorders encompass tension-type headaches, migraines, stroke, Alzheimer's disease (AD), and Parkinson's disease (PD), with tension-type headaches impacting around 2 billion people worldwide [2].

Neurological disorders do not occur uniformly; their distribution is significantly affected by age, sex, and geographical location. In 2019, there were approximately 805.17 million reported cases worldwide, showing a significant rise in age-standardized incidence rates for conditions such as PD and idiopathic epilepsy [1, 3]. Furthermore, the impact of these disorders is unevenly experienced in low- and middle-income countries, where access to treatment remains restricted. Since 1990, the disability-adjusted life years (DALYs) linked to neurological conditions have risen by 18%, highlighting the critical necessity for effective diagnostic and therapeutic approaches [1, 4].

M. R. Alam (✉)
Department of Computer Science, Faculty of Science and Information Technology, American International University-Bangladesh, Kuratoli, Bangladesh

A. Al Abid
Department of Software Engineering, Faculty of Science and Information Technology, Daffodil International University-Bangladesh, Daffodil Smart City, Birulia, Bangladesh
e-mail: abid35-757@diu.edu.bd

M. M. Hassan et al. (eds.), *Graph Neural Networks for Neurological Disorders*,
https://doi.org/10.1007/978-3-032-04315-3_2

Timely diagnosis and intervention play a vital role in effectively managing neurological disorders. Early detection helps to improve outcomes by allowing targeted treatments that stop disease onset or lower symptoms. Still, conventional diagnostic techniques occasionally fall short since they depend on subjective evaluations and the complicated nature of neurological symptoms. This discrepancy highlights the need of creative ideas to increase diagnosis accuracy and therapy efficacy.

ML, a subset of AI, is now an evolving agent in the healthcare sector, especially within neurology. By using large datasets and sophisticated algorithms, ML enables doctors to recognize trends that could go undetected with conventional analysis. Applications of ML in healthcare cover a spectrum from diagnostic imaging to predictive analytics to individualized treatment.

In neurology, ML algorithms are increasingly employed to analyze neuroimaging data, including magnetic resonance imaging (MRI) and computed tomography (CT) scans, to identify abnormalities linked to different neurological conditions. From brain scans, studies have shown that ML models can precisely detect characteristics indicating AD or multiple sclerosis [3]. Early treatments and improved diagnosis accuracy these days greatly improve patient outcomes. In neurology, ML-driven predictive analytics is vital since it analyzes medical history and demographic data to help identify patients at great risk for particular diseases.

ML-driven predictive analytics is crucial in neurology, as it helps identify patients at high risk for specific conditions by analyzing their medical history and demographic factors. By analyzing trends found from electronic health records, for example, algorithms can predict the probability of stroke or seizure recurrence [5]. This capacity helps medical professionals to apply preventive actions on their own initiative.

Furthermore, personalized medicine depends much on ML since it enables tailored treatment plans to fit for each patient's profile. By means of data obtained from clinical trials and real-world evidence, ML models can help to discover the most efficient treatments catered for patient populations. This approach shows particularly helpful in treating chronic neurological diseases since individual treatment responses vary greatly.

Although it has great potential, including ML into clinical practice faces many challenges. Data privacy issues, the need of extensively annotated datasets to train algorithms, and the value of cooperation between doctors and data scientists need careful attention. Moreover, it is crucial to guarantee that ML technologies are transparent and understandable to inspire general acceptance and create clinician confidence.

Emphasizing its capacity to provide individualized therapies, increase diagnosis accuracy, and forecast illness progression, this study investigates the transforming possibilities of ML in neurology. However, effective integration of ML into clinical practice calls for resolving major ethical, technical, and legal issues. The key outcomes for the study consist in:

- ***Personalized Treatments:*** ML improves patient-specific treatment by enabling customized approaches for neurological diseases.

- ***Improved Diagnosis:*** To increase diagnosis accuracy, ML systems examine complicated neurological data including neuroimaging, genetic information, and behavioral features.
- ***Disease Progression Prediction:*** ML models help to forecast the course of neurological disorders, therefore facilitating early intervention and improved control.
- ***Ethical and Technological Considerations:*** The study highlights critical concerns, including data bias, patient information security, and the risks of over-reliance on "black-box" systems.
- ***Regulatory Adaptation:*** Modern regulatory systems are needed to guarantee rigorous clinical testing and validation of AI-driven solutions.
- ***Transparency and Human Oversight:*** Maintaining trust and responsibility in decision-making processes depends on interpretable ML systems and human supervision.
- ***Interdisciplinary Challenges:*** The research identifies four key challenges for realizing ML's full potential in neurology:
  - Enhancing data quality and standards.
  - Developing interpretable ML architectures.
  - Addressing ethical implications.
  - Fostering collaboration among stakeholders.

These outcomes underscore the promise of ML in revolutionizing neurology care while addressing the barriers to its effective and ethical implementation.

In the following sections, these topics have been discussed: Section 2.2 provides an overview of key studies and advancements in the application of machine learning (ML) to neurology, highlighting foundational research, emerging trends, and gaps in the current knowledge base. Section 2.3 explores the types of data used in neurology, such as neuroimaging, electrophysiological recordings, and clinical data, while addressing challenges related to data quality, availability, and preprocessing. Section 2.4 focuses on the core ML algorithms, including supervised, unsupervised, and reinforcement learning, and their relevance to solving neurological problems. Section 2.5 delves into practical applications of ML in neurology, such as disease diagnosis, prognosis prediction, treatment personalization, and drug discovery. Section 4.6 discusses the metrics used to assess the performance of ML models in neurology, including accuracy, precision, recall, sensitivity, specificity, and area under the curve (AUC), among others. Section 2.7 addresses the ethical challenges of using ML in neurology, including issues of data privacy, algorithmic bias, and the need for transparency and accountability in decision-making. Section 2.8 outlines emerging trends and research opportunities, such as explainable AI, federated learning, and multimodal data integration, to advance the field of ML in neurology. Section 4.9 summarizes the key insights from the chapter, emphasizing the transformative potential of ML in neurology while underscoring the importance of addressing technical, ethical, and practical challenges for its successful implementation.

2.2 Literature Review

2.2.1 Overview of ML Applications in Neurology

ML has emerged as a transformative tool in neurology, offering innovative methods for the diagnosis and treatment of neurological illnesses [48]. By including ML techniques into clinical practice, it becomes easier to investigate large databases, including electronic health records, genetic profiles, and neuroimaging scans, to find trends that could remain latent using traditional methods. For example, ML models have been employed in neuroimaging-based predictions of PD, attaining significant accuracy through the integration of multimodal data, including neuroimaging, plasma biomarkers, and clinical assessments [6]. Furthermore, advanced techniques such as graph neural networks (GNNs) and deep graph convolutional neural networks (DGCNNs) have shown their effectiveness in analyzing functional neuroimaging data for disorders like schizophrenia, achieving sensitivity and accuracy that exceed traditional methods [7]. Moreover, hybrid quantum ML pipelines, which integrate classical convolutional neural networks (CNNs) with Quantum Support Vector Machines (QSVM), have enhanced the classification of dementia severity [8]. Research has utilized ML models to distinguish between bipolar disorder and major depressive disorder by analyzing brain network metrics, revealing distinct neuroimaging biomarkers for these disorders [9]. Emphasizing the need to address inequalities and biases in dataset representation to guarantee model generalizability and inclusivity, ethical issues of using neuroimaging resources for ML have been underlined [10]. The achievements underscore the capacity of ML to augment diagnostic precision and facilitate early identification of ailments like AD and PD, consequently markedly improving patient outcomes [11].

ML methods, for instance, are being employed in clinical decision support systems (CDSS) that analyze extensive neurological and medical data to help practitioners reach informed decisions. These systems have demonstrated active contributions to tailoring treatment approaches according to individual patient data and predicting disease trajectory. The advanced clinical decision support (CDS) framework proposed by Mukherjee et al. [12] amalgamates a multi-agent multiple feature evaluation approach (MFEA), the synthetic minority oversampling technique (SMOTE), and artificial neural networks (ANN), achieving a remarkable accuracy rate of 96.3% for the early diagnosis of PD. In a similar vein, the longitudinal healthcare analytics system created by Owusu et al. [13] incorporates deep learning models applied to clinical and neuroimaging data to forecast the progression of AD. This decision support system, which operates in two stages, utilizes transfer learning, 3D convolutional neural networks (3D-CNN), and occlusion maps to improve prediction accuracy and enhance model interpretability for medical practitioners. Despite the significant advancements demonstrated by ML, as noted by Masood et al. [14], the challenges in integrating CDSS with picture archiving and communication systems (PACS) continue to hinder its widespread clinical adoption. Furthermore, the ML-based CDSS developed by Tang et al. [15] for neonatal sepsis showed more than 80% prediction accuracy for β-lactam antibiotic dosing, greatly enhancing target attainment probabilities in comparison to guideline-recommended doses. This research taken together show the transforming power of AI-driven CDSS in raising diagnosis accuracy, streamlining treatment approaches, and so improving patient outcomes.

Given the enormous volume of data generated by continuous patient monitoring, the use of ML is especially important within intensive care units (ICUs). Algorithms created to analyze physiological waveforms and electroencephalogram (EEG) data can greatly reduce the workload for healthcare providers by automating the analysis process. Tang et al. [15] created ML models, including logistic regression (LR) and random forest (RF), aimed at predicting neurological outcomes in critically ill patients suffering from hemorrhagic and ischemic stroke. With AUC values of 0.887 and 0.867, respectively, their efforts displayed remarkable accuracy. By including SHapley Additive exPlanations (SHAP), these models' interpretability was much enhanced since proper clinical decision-making depends on important predictors like GCS score, APS III score, and SOFA score. In a similar vein, Wang et al. [16] utilized a multilayer perceptron neural network to forecast the incidence of ICU-acquired weakness (ICU-AW), attaining impressive results with an AUC of 0.941. Important risk factors—including the length of ICU stay, the duration of mechanical ventilation, and higher dosages of sedatives and vasopressors—which can direct early interventions—were underlined in the study. Through precise and prompt treatments, these developments speed the diagnosis process and improve patient management.

Additionally, natural language processing (NLP) advancements allow therapeutically pertinent details to be processed from the unstructured text in EHRs. By providing enhanced capacity to retrieve and analyze non-sorted text data in great detail and in large quantities, such technology will carry great benefits from simplifying health care processes and contributing to the research projects [11]. For diagnosis, phenotyping, and prognosis of neurological diseases, NLP has shown success. Lefkovitz et al. [17] examined 41 research and underlined how well NLP may improve diagnosis accuracy in settings where neurological knowledge is rare. Likewise, Kim et al. [18] evaluated ChatGPT-4's ability to answer queries on epilepsy and found it to be a good source of reliable medical knowledge. Using NLP, Lo Barco et al. [19] extracted phenotypic ideas from narrative health data, therefore enabling the thorough mapping of the natural history of uncommon disorders such as Dravet syndrome. To extract headache frequency from clinical notes, Chiang et al. [20] also developed and polished NLP models including the GPT-2 frameworks, so reaching a startling accuracy of 92%. Particularly in the administration of unstructured medical data, the developments reveal the transformational possibilities of NLP in clinical and research environments.

2.2.2 *Contributions to Personalized Medicine*

With the most focused view on to personalized medicine in neurology, ML has come to the potential to offer insights in which treatment strategies could be individualized per patient, each bearing its characteristics and needs. Algorithms analyzing various data sources—like genetic information, medical history, and real-time physiological data—with considerations of each patient's characteristics would enable healthcare providers to design better treatment programs. For example, cluster analysis driven by ML has been employed to pinpoint unique patient phenotypes, including individuals with degenerative cervical myelopathy (DCM), through the examination of compiled clinical data. This approach revealed four different characteristics, each marked by different functional recovery trajectories and predictors, including frailty score and symptom duration, thereby helping to develop treatment plans catered to certain phenotypes [21].

Models based on ML have been utilized to forecast treatment responses in individuals suffering from chronic neurological disorders. Predictive analytics using modular chronic illness models, for example, have accurately forecast obesity and chronic obstructive pulmonary disease, so highlighting the important link between lifestyle, environmental effects, and health. This approach shows how ML may generate economic models with less features, hence improving their relevance in practical clinical settings [22]. These models let doctors identify the most likely successful treatments for each patient, therefore improving therapeutic results and lowering side effects. Moreover, reinforcement learning-based rehabilitation techniques have shown how able they are to change therapy plans based on the particular responses of every patient during their recovery process [23].

The integration of ML into drug discovery procedures is essential for the progression of customized medicine. ML systems can identify possible therapeutic targets and project the reactions of different patient populations to novel drugs by means of the analysis of large datasets obtained from clinical trials and genomic investigations. This method speeds up new treatment creation and assures that new therapies correspond more closely to the biological characteristics of specific patient populations. An example includes the application of ML to the analysis of disease trajectories and treatment response variation through real-world datasets and omics data [48]. This application enhances disease prediction and patient stratification through methods such as clustering and tree-based ensemble techniques [24]. In the realm of neurology, developments in AI have greatly improved early illness diagnosis, neuroimaging, and the personalizing of treatments for neurodegenerative disorders, so highlighting greater diagnosis accuracy and treatment strategy optimization [25]. Moreover, ML has transformed preclinical drug discovery by speeding up virtual screening, discovering new bioactive molecules, and enhancing predictive models for drug properties [26]. In the realm of neurodegenerative diseases, the integration of ML with molecular docking has led to the identification of ligands with therapeutic potential, the proposal of early biomarkers, and the development of AI-designed molecules for conditions such as AD and PD [51]. This emphasizes the need to include computer techniques in the process of developing drugs [27].

2.2.3 Challenges and the Importance of Interdisciplinary Collaboration in Machine Learning for Neurology

Applications of ML in neurology result in personalized medicine, which lets treatment plans be tailored to every patient's particular requirement and trait. By using algorithms that examine many data sources, including genetic information, medical history, and real-time physiological data, healthcare providers can develop more efficient treatment regimens considering the particular traits of every patient. For example, cluster analysis driven by ML has been employed to pinpoint unique patient phenotypes, including individuals with degenerative cervical myelopathy (DCM), through the examination of compiled clinical data. This approach revealed four different phenotypes, each marked by different functional recovery trajectories and predictors, including frailty score and symptom duration, thereby helping to develop treatment plans catered to certain phenotypes [21].

Models based on ML have been utilized to forecast treatment responses in individuals suffering from chronic neurological disorders. Predictive analytics using modular chronic illness models, for example, have accurately forecast outcomes like obesity and chronic obstructive pulmonary disease, so highlighting the important link between lifestyle, environmental factors, and health. This approach shows how ML may generate economic models with less features, hence improving their relevance in practical clinical settings [22]. These models let doctors identify the most likely efficient treatments for each patient, therefore improving therapeutic results and lowering side effects. Moreover, reinforcement learning-based rehabilitation techniques have shown how able they are to change therapy plans based on the particular responses of every patient during their recovery process [23].

The incorporation of ML into drug discovery processes plays a crucial role in advancing personalized medicine. ML systems can identify possible therapeutic targets and project the reactions of different patient populations to novel drugs by means of the analysis of large datasets obtained from clinical trials and genomic investigations. This approach guarantees that novel treatments are more precisely fit to the biological traits of particular patient populations in addition to accelerating their development. Using real-world datasets and omics data, ML has, for example, efficiently examined illness trajectories and the heterogeneity in therapy responses. This application enhances disease prediction and patient stratification through methods such as clustering and tree-based ensemble techniques [25]. In a broad set of neuroscientific applications, AI developments have contributed greatly towards improved early diagnosis of diseases, neuroimaging, and the personalization of neurodegenerative diseases treatment, attracting numerous manufacturers due to improved diagnostic accuracy and optimization of treatment strategies [26]. Moreover, ML has transformed preclinical drug discovery by speeding up virtual screening, discovering new bioactive molecules, and enhancing predictive models for drug properties [27]. In the realm of neurodegenerative diseases, the integration of ML with molecular docking has led to the identification of ligands with therapeutic potential, the proposal of early biomarkers, and the development of AI-designed molecules for conditions such as Alzheimer's and PD. This highlights the significance of incorporating computational methods in the drug development process [28, 51].

Finally, the literature review emphasizes the transformative power of ML in neurology, particularly in enhancing diagnostic accuracy, customizing treatment plans, and improving patient outcomes. Integrating ML algorithms into clinical practice enables the analysis of vast datasets, uncovering patterns that traditional methods may overlook. Notable advancements include the use of neuroimaging data for early detection of diseases such as Parkinson's and Alzheimer's, the development of intelligent clinical decision support systems (CDSS), and the application of natural language processing (NLP) to extract insights from unstructured health records. Furthermore, ML facilitates personalized medicine by tailoring treatment plans to individual patient characteristics, thereby optimizing therapeutic outcomes. Despite these advancements, challenges such as data representation biases and the need for model interpretability remain, necessitating multidisciplinary collaboration to address these issues. A summary of key ML applications in neurology, including the data types, methodologies, and outcomes, is provided in Table 2.1. Overall, the findings underscore the potential of ML to revolutionize neurology and pave the way for more personalized and effective patient care.

2.3 Data in Neurology

ML has demonstrated significant promise in transforming the field of neurology, especially through its use across diverse data sources in clinical trials, neuroimaging, and genomics. The several kinds of data used in neurological research are discussed in this part, together with the necessary preprocessing methods for preserving data integrity and the difficulties in managing neurological datasets.

2.3.1 Categories of Data Sources

In neurology, the development and improvement of ML models depend on many kinds of data. Included are clinical data, neuroimaging data, and genetic data; each contributes in different ways to many aspects of neurological research and clinical decision-making.

- **Clinical Data**

Understanding patient backgrounds and medical conditions depends much on clinical data—including patient histories and electronic medical records (EMRs). Miller et al. (2023) discuss in their paper on ML applications in clinical trials that ML methodologies have the potential to greatly improve clinical trial processes, such as patient recruitment and remote monitoring [29]. Clinical data play a very

Table 2.1 Summary of machine learning applications in neurology: data, methodologies, and outcomes

Author Name	Data	Methodology	Outcomes
Zhu et al. [6]	Multimodal neuroimaging, plasma biomarkers, clinical assessments	Multimodal ML integration (RF, SVM)	Notable accuracy in PD prediction
Sunil et al. [7]	Functional neuroimaging	GNNs and DGCNNs	Sensitivity and accuracy surpassing traditional methods in schizophrenia
Chen et al. [8]	High-dimensional neuroimaging	Hybrid quantum ML (CNN + QSVM)	Improved dementia severity classification
Huang et al. [9]	Structural connectome data	ML models (unspecified)	Unique neuroimaging biomarkers for bipolar vs. MDD distinction
Lock et al. [10]	Neuroimaging repositories	Ethical analysis of dataset representation and biases	Highlighted disparities in dataset inclusivity and generalizability
Mukherjee et al. [12]	Clinical data for PD	Multi-agent MFEA + SMOTE + ANN	96.3% accuracy in early detection of PD
Owusu et al. [13]	Clinical + neuroimaging data	3D-CNN + transfer learning + occlusion maps	Early detection and monitoring of AD progression; enhanced interpretability
Masood et al. [14]	Multimodal datasets: Lung (LIDC-IDRI), brain (BRATS), skin (HAM10000), and others.	Systematic review of ML-based CDSS across cancers and neurological diseases. Methods: SVM, DT, RF, CNN, federated learning	Highlighted challenges in CDSS adoption: **Integration barriers** with PACS and EHR systems. Need for standardized data formats and federated learning to address interoperability
Tang et al. [15]	Neonatal sepsis data	ML-based CDSS	>80% accuracy in β-lactam antibiotic dosing optimization
Wang et al. [16]	ICU patient data	Multilayer perceptron	AUC 0.941 for ICU-AW prediction; identified risk factors
Lefkovitz et al. [17]	Unstructured EHR text	NLP for diagnostic phenotyping	Potential for diagnostic accuracy in limited-expertise settings
Kim et al. [18]	Epilepsy-related questions	ChatGPT-4 evaluation	Reliable medical information retrieval for epilepsy
Barco et al. [19]	Narrative EHRs (Dravet syndrome)	NLP for rare disease phenotyping	The mapped natural history of Dravet syndrome
Chiang et al. [20]	Clinical notes (headache frequency)	GPT-2 fine-tuning	92% accuracy in headache frequency extraction
Wei et al. [28]	Physiological waveforms, EEG data	LR, RF + SHAP explanations	AUC 0.887 (LR) and 0.867 (RF) for stroke outcome prediction; key predictors identified

important role in constructing a comprehensive patient profile, which is crucial to tailoring treatment and interventions around neurology. Also, Gourie-Devi et al. underscore the importance of clinical data for understanding the prevalence of neurological diseases in different populations [4].

- **Neuroimaging Data**

Neurology uses neuroimaging data such as MRI, CT scans, and EEG to examine brain structure, function, and any abnormalities present. Incorporating ML into neuroimaging has yielded encouraging outcomes, as highlighted by Stumpo et al. [30] and Davatzikos [31] who utilize ML and deep learning methods to uncover intricate patterns within neuroimaging data [30, 31]. These programs are important since they can identify neurological disorders such as brain tumors and AD, therefore guiding treatment. Moreover, underscored by Yue Wang et al. is the important impact of neurologic diseases in Asian nations based on neuroimaging data to detect common diseases and related risks [5].

- **Genetic Information and Indicators**

Genetic data and biomarkers provide insights into the underlying biological mechanisms of neurological diseases. Li Shen and P. Thompson (2020) emphasize the developing area of brain imaging genomics, which combines genetic data with brain imaging to enhance our comprehension of neurological disorders [32]. Researchers can create more precise models for predicting disease progression and patient outcomes by integrating imaging data with genetic and clinical data. Yi Huang et al. emphasize the amalgamation of biomarkers with imaging and clinical data, employing sophisticated algorithms to evaluate disease progression and therapy effectiveness [4].

2.3.2 Data Preparation

Data preparation is the preliminary step in which raw data is prepared for a modeling process in a ML framework. Because of the complex and variable nature of the compositions involved in the data, neurology requires good preprocessing techniques to ensure the achievement of good models.

- **The Significance of Data Cleaning and Normalization**

Data cleaning refers to the actual elimination of noise or inconsistencies in the process of creating clinical or neuroimaging information; in this case, it features a more specific application: the very large piles of clinical or neuroimaging data. Thomas et al. [33] address the difficulty of missing data in ML research targeted at brain diseases [33]. Missing values can seriously affect model performance; hence, imputation is used to solve this problem. Especially in neuroimaging, data normalization is quite important since variations in scanner type and acquisition technique might lead to data discrepancies. Standardizing these data points helps us to ensure that the ML model's input is constant and unbiased [41].

- **Techniques for Feature Extraction and Selection**

Feature extraction is the process of identifying, from complex datasets including neuroimaging and genetic data, the most relevant information. The application of radiomics for feature extraction from medical pictures is under debate in Stumpo et al. [30], stressing it's possible to enhance image characterization and prognosis prediction [30]. Methods for feature selection help to identify the most significant factors improving predictive power, therefore reducing dimensionality and increasing the efficiency of the model. Furthermore, the studies conducted by B.K. MacDonald et al. [34] underscore the necessity of meticulously selecting pertinent variables to guarantee reliable predictions in neurological research [34].

2.3.3 Difficulties in Neurological Data

ML has great promise in neurology; nevertheless, processing neurological data presents certain difficulties. The key difficulties include the quality, completeness, and bias in the data, all of which significantly influence the performance of ML models [41].

- **Concerns Regarding Data Quality**

Using clinical and neuroimaging data for ML is a major difficulty, given the natural diversity in data quality. Particularly with regard to data consistency and quality in the field of clinical neurology, Smith et al. [35] underline the difficulties that medical practitioners face as ML applications develop [35]. Inaccuracies in patient records, variations in imaging protocols, and discrepancies in diagnostic labels can all lead to suboptimal model training. The research conducted by Yi Huang et al. [4] highlights the importance of stringent data quality assurance in neurological datasets to guarantee precise modeling [4].

- **Absence of Data**

Missing data is another often-occurring difficulty, particularly in studies on rare neurological diseases and therapeutic trials. Missing data can seriously affect the dependability of ML models, claims R. Thomas et al. (2020) [33]. In clinical environments, the difficulty is usually increased by the lack of some information, which could result from technical restrictions, privacy issues, or resource limits. Solving these problems depends critically on advanced imputation methods and models designed for handling partial data [33].

- **Partiality in Information**

Particularly when creating models from clinical statistics that might not fairly represent the variety of patient populations, bias presents a significant obstacle. One can add bias in several ways, including through sample techniques or cultural influences on treatment choices. Miller et al. (2023) claim that including ML in healthcare operations causes different administrative and legal difficulties. Efforts meant to reduce prejudice must be addressed if we are to ensure models behave equally across several demographic groupings [29].

2.4 Algorithms in Machine Learning

ML has transformed how we analyze and interpret intricate neurological datasets, laying the groundwork for personalized treatments, enhanced diagnostics, and predictive modeling. Along with a thorough study, this part explores the foundations of ML, its several forms, often-used algorithms, and considerations to be taken into account while deciding the most appropriate technique.

2.4.1 *What Is Machine Learning?*

ML refers to a subset of AI that enables systems to learn and improve from experience without being explicitly programmed. It involves the development of algorithms that can analyze and interpret data, allowing computers to make decisions or predictions based on that information. By utilizing statistical techniques, ML models can identify patterns and trends within large datasets [52], leading to more accurate outcomes in various applications, from image recognition to natural language.

- ***Definition and Core Concept of Machine Learning***
  - ML is a segment of AI that allows computer systems to acquire knowledge and enhance their performance based on experience without explicit programming.
  - In contrast to traditional systems, ML employs data-driven techniques to reveal patterns and relationships, proving to be exceptionally effective in handling a variety of datasets, including clinical records, neuroimaging scans, and genomic sequences.
- ***Neurology Applications***
  - By revealing complex linkages in data, ML systems enable revolutionary uses in neurology like medication optimization and illness progression modeling [4].
  - The ability of ML to adapt to new data presents a considerable benefit, enabling models to independently refresh themselves with incoming patient information, which in turn enhances predictive accuracy as time progresses [35].
- ***Handling Complex Data***
  - Given the complex character of neurological data, ML shows remarkable capacity in resolving non-linear correlations seen in high-dimensional datasets. This is a crucial need in neurological research [6].

  - Traditional neuroimaging programming depends on established criteria to identify abnormalities, while ML algorithms, such as CNNs, independently learn and recognize features most indicative of neurological conditions [32].

- ***Comparison with Traditional Programming***
  - In traditional programming, rules and logic are explicitly defined to address problems, whereas ML algorithms "learn" from data to produce predictions or make decisions.
  - The self-learning capability of ML is especially beneficial in clinical neurology, where datasets tend to be heterogeneous and are constantly evolving.

- ***Flexibility in Clinical Workflows***
  - ML is progressively utilized to meet the changing demands of clinical workflows, including integrating patient records from various sources for thorough analysis [29].
  - This flexibility guarantees that ML models remain pertinent and efficient in ever-changing clinical settings.

2.4.2 Categories of Machine Learning

- ***Supervised Learning***

Under supervised learning, models are trained using labeled data under the direction of input-output pairs. In neurology, supervised learning finds extensive use, especially in disease classification and the prediction of progression. Stumpo et al. demonstrate how labeled neuroimaging datasets can be applied in training ML models for the early detection of AD, leveraging patterns identified in MRI and PET scans [30]. Additionally, R. Thomas et al. highlight the importance of supervised learning in predicting treatment outcomes for stroke patients with historical clinical data [33]. In the realm of epilepsy research, various supervised learning techniques, such as support vector machines and decision trees, have been employed. Zhu et al. explored the improvement of diagnostic accuracy in distinguishing seizure types by classifying EEG data with supervised techniques [6].

- ***Unsupervised Learning***

Unsupervised learning reveals latent patterns and structures without any predefined outputs on data free of labels. In the field of neurology especially, this is quite helpful for exploratory data analysis. Using MRI feature analysis, Yue Wang et al. underline how clustering techniques have been used to partition patient groups and help to identify subtypes in disorders including multiple sclerosis [5]. In the field of brain imaging genomics, unsupervised learning has been essential in uncovering gene-expression patterns associated with differences in brain structure [32]. Dimensionality reduction techniques, such as principal component analysis (PCA), have been employed to simplify complex neuroimaging datasets for further analysis [34].

- ***Reinforcement Learning (RL)***

Training an agent to perform a series of decisions aiming at optimizing cumulative rewards in a simulated environment is the essence of RL. The potential exists to enhance treatment strategies for neurological disorders. Miller et al. highlight the potential of RL in simulating therapeutic pathways for conditions like PD, enabling models to propose the most effective interventions [29]. RL has been used in rehabilitation to create robotic-assisted therapy schedules, especially for stroke victims. Based on patient progress, the agent real-time changes therapeutic parameters, so improving recovery results [10]. Although RL is still developing in the field of neurology, its ability to enhance long-term therapy plans is rather important.

2.4.3 Overview of Common Algorithms

- ***Decision Trees***

Simple yet effective models that partition data into subsets based on feature values help to improve the interpretability of decision trees by themselves [42]. In Fig. 2.1, the decision-making process of a decision tree algorithm is shown. The tree consists of one root node and multiple child node (decision nodes), each of which enables the algorithm to plan based on the given criteria. Colin M. Smith and colleagues discuss their method of examining patient histories with an eye on early neurological disease diagnosis [32]. Their capacity to handle numerical and categorical data makes them very useful for jobs such as epilepsy classification or stroke recurrence prediction.

- ***Random Forests***

Random forests, which are an ensemble method utilizing decision trees, enhance both accuracy and robustness through the combination of outputs from multiple trees [47]. In

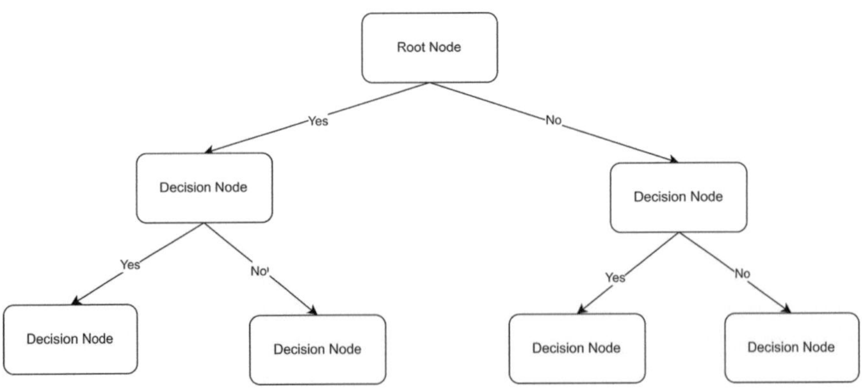

Fig. 2.1 Decision tree architecture

Fig. 2.2, the working principle of the random forest algorithm has been shown. The figure shows that multiple trees are used, and the average of their prediction is taken as the final output. MacDonald et al. [34] illustrate their application in the analysis of noisy clinical datasets, especially in predicting outcomes for patients with traumatic brain injury. Random forests excel at managing missing data and mitigating overfitting, a frequent challenge encountered in neuroimaging studies.

- ***Support Vector Machines (SVMs)***

SVMs are well-suited for neuroimaging due to their ability to classify high-dimensional data effectively. As shown in Fig. 2.3, SVM identifies the optimal hyperplane (wx + b = 0) that maximizes the margin between two classes, with support vectors defining the boundary. The provided image illustrates this concept, showing margin boundaries (wx + b = ±1) and the role of support vectors in classification. SVMs handle non-linearly separable data using kernel functions and balance misclassification with a regularization parameter (C). Studies by Stumpo et al. [30] and Zhu et al. [6] highlight SVM's effectiveness in distinguishing AD from normal aging and classifying EEG signals for epilepsy diagnosis.

- ***Neural Networks***

- Modern machines consist fundamentally in neural networks, providing robust tools for the analysis of intricate datasets. Various architectures are tailored for specific applications within the field of neurology:
  - ***Perceptron:*** Early neural networks, such as the perceptron, were primarily used for binary classification tasks. The perceptron, a foundational model

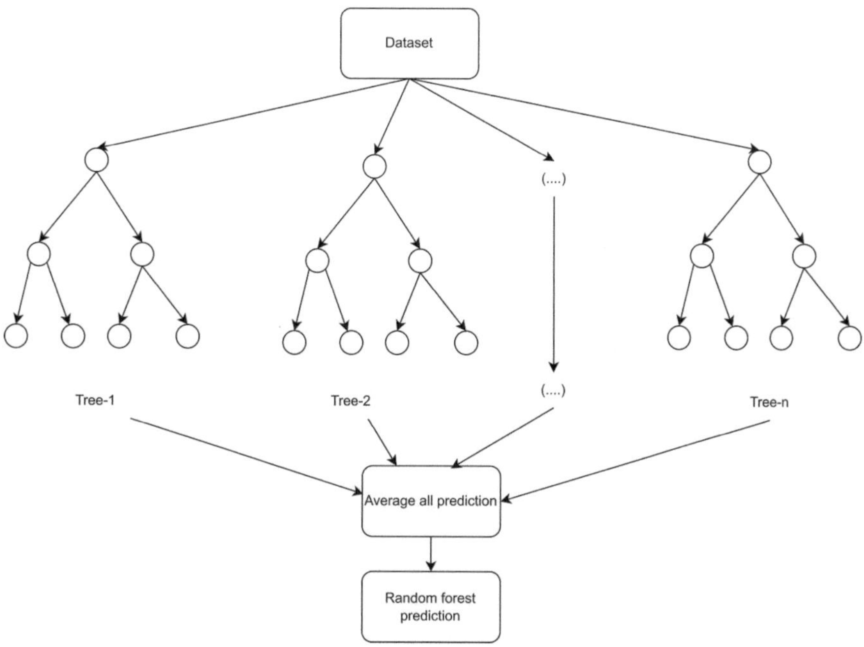

Fig. 2.2 Random forest architecture

Fig. 2.3 SVM architecture

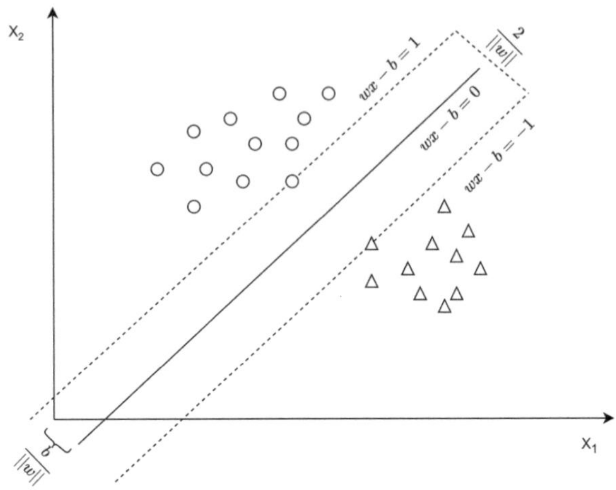

in neural networks, operates by computing a weighted sum of inputs, applying an activation function (often a step function), and producing a binary output. The architecture of the perceptron model, including its inputs, weights, and activation process, is explained in Fig. 2.4. While effective for linearly separable data, its limitations in handling non-linear problems were notable. As Yi Huang et al. emphasize, despite their restricted scope, perceptron's laid the groundwork for more complex architectures, paving the way for advanced models like multi-layer perceptrons (MLPs) and deep neural networks [4]. This evolution highlights the perceptron's role as a critical steppingstone in the development of modern machine learning.

- ***Convolutional Neural Networks (CNNs):*** Particularly suited for handling image data, CNNs have become popular in neuroimaging, especially for detecting brain tumors in MRI scans with high diagnostic accuracy [30]. Figure 2.5 illustrates a CNN architecture used for classification tasks. The process begins with an input image, where small regions are analyzed by convolutional filters to detect patterns. The feature extraction stage consists of convolution layers, which identify important features like edges and textures, followed by pooling layers, which reduce spatial dimensions while retaining essential information. Finally, in the classification stage, extracted features are passed through a fully connected layer, leading to an output layer that makes a final decision (e.g., YES/NO). This CNN pipeline plays a crucial role in neuroimaging applications, aiding in accurate and automated medical diagnoses.

- ***Recurrent Neural Networks (RNNs):*** RNNs are specifically crafted for the analysis of sequential data, making them particularly suitable for time-series applications such as EEG analysis. Their ability to retain information from previous time steps allows them to capture temporal dependencies, which is crucial for

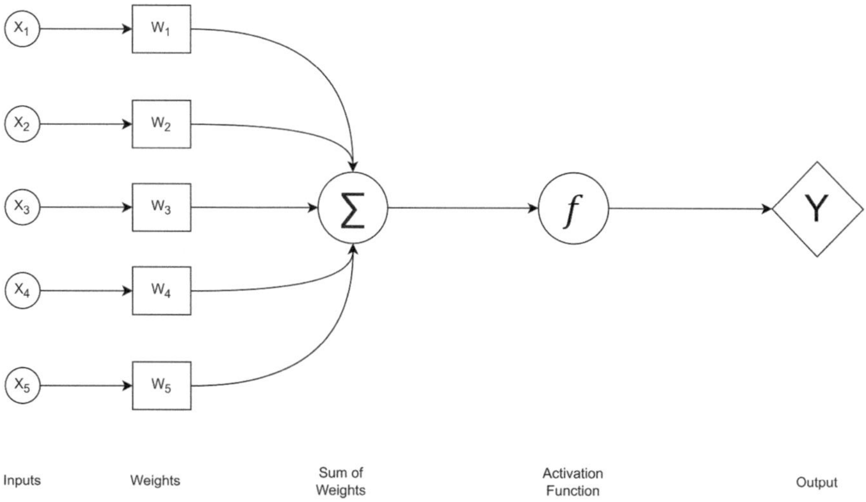

Fig. 2.4 Perceptron architecture

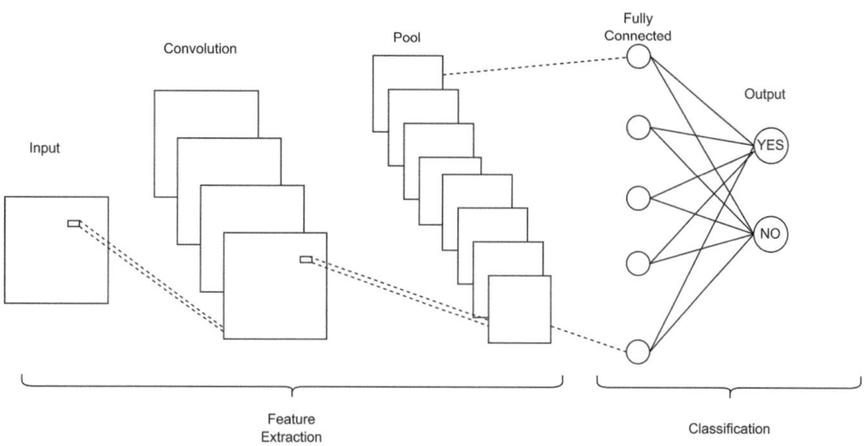

Fig. 2.5 Convolution neural network (CNN) Architecture

tasks like predicting seizure onset, as explored by Zhu et al. [6]. Figure 2.6 illustrates the architecture of a RNN, consisting of an input layer, multiple hidden layers, and an output layer for binary classification. Unlike traditional feedforward networks, RNNs incorporate recurrent connections within their hidden layers, enabling the model to process sequential information effectively. Each hidden layer passes information to the next time step, allowing the network to learn complex temporal patterns. The final output layer determines the classification outcome (e.g., seizure or no seizure). This architecture has been widely used in neuroimaging and EEG-based applications, improving the accuracy of time-dependent medical diagnoses.

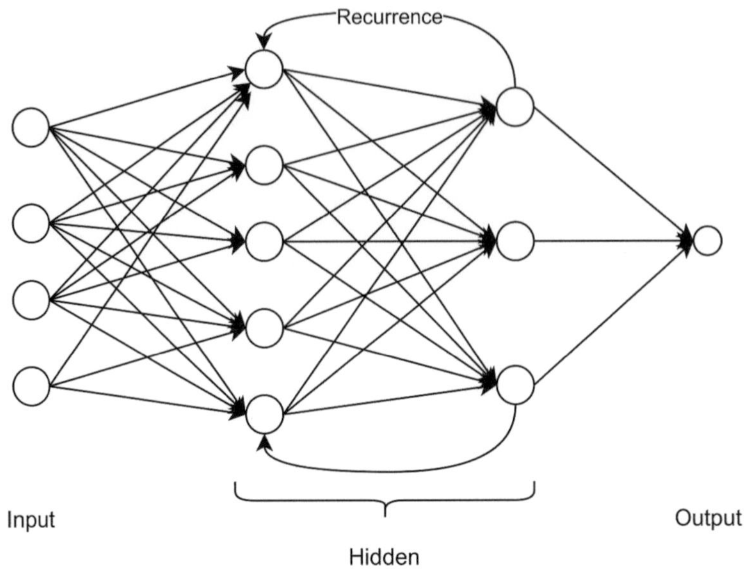

Fig. 2.6 Recurrent neural network (RNN) architecture

– ***Graph Neural Networks (GNNs):*** GNNs analyze relational data, such as brain connectivity networks, making them highly effective for studying complex neural interactions. Using GNNs, Yue Wang et al. [5] investigate abnormalities in functional connectivity linked with autism and schizophrenia. Figure 2.7 illustrates the architecture of a GNN, which consists of an input layer, multiple hidden layers, and an output layer. Unlike traditional neural networks, GNNs operate on graph-structured data, where nodes represent entities (e.g., brain regions), and edges denote relationships (e.g., functional connectivity). The hidden layers apply iterative message passing, aggregating information from neighboring nodes to learn meaningful graph representations. The output layer ultimately classifies or predicts outcomes based on the learned features. This capability makes GNNs particularly valuable for analyzing the intricate connectivity patterns in the human brain, aiding in the identification of neurological disorders.

2.4.4 Choosing the Right Algorithm

Factors to Consider

The success of ML implementations in the field of neurology depends on the chosen method [42]. Important factors to consider are:

- ***Data Type:*** Various types of data necessitate customized algorithms. For example, CNNs excel in handling neuroimaging data, whereas RNNs are more appropriate for analyzing EEG time-series data [30].

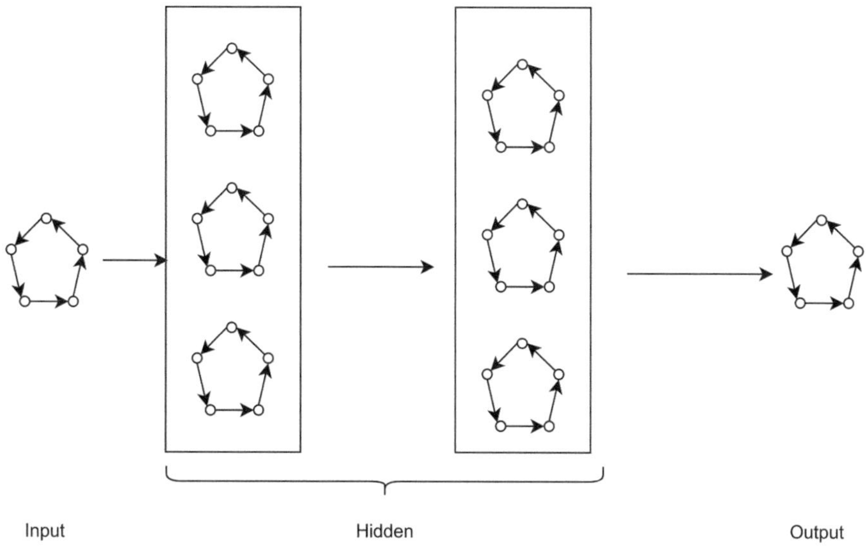

Fig. 2.7 Graph neural network (GNN) architecture

- ***Problem Complexity:*** While more complex problems including multi-modal data fusion call for the employment of neural networks or ensemble approaches, for simple classification jobs decision trees or SVMs may suffice [35].
- ***Interpretability:*** The value of interpretability determines whether clinician confidence in therapeutic applications is justified. Colin M. Smith et al. [35] underline the need of using interpretable models, including decision trees, for activities requiring openness.

Thoughtfully matching the algorithm choice to the dataset and research objectives helps researchers improve the effectiveness of ML applications in neurology. This approach enhances not only model performance but also encourages more general acceptance in clinical practice.

2.5 Applications of Machine Learning in Neurology

ML has had a major impact on neurology since it has promoted development in disease prediction, imaging analysis, individualized therapy, and disease management. Emphasizing case studies, approaches, and lessons learned from current research, this part investigates the several uses of ML within the discipline of neurology. By including ML methods, therapy efficacy and diagnostic accuracy have been raised as well as new paths for knowledge and control of neurological diseases.

2.5.1 *Disease Prediction and Classification*

Predicting and classifying neurological diseases has demonstrated remarkable success for machine learning (ML). Utilizing diverse datasets, including genetic, neuroimaging, and clinical data, ML models have enabled earlier diagnosis and more precise treatments. As illustrated in Fig. 2.8, the disease prediction workflow begins with data collection, encompassing image data, clinical information, and wearable data. This is followed by the Data Preprocessing stage, which involves cleaning, normalization, and feature engineering. The next phase is Model Selection, where techniques such as supervised, unsupervised, semi-supervised, and transfer learning are employed. Subsequently, the Model Prediction phase involves applying the trained model to a test dataset, and finally, the Model Evaluation phase assesses performance using metrics like accuracy, precision/recall, and F1 score. Case studies covering various neurological disorders are explored in this area, highlighting the transformative power of ML in disease prediction and classification [40].

- **Case Studies on Alzheimer's Disease**

  AD stands out as one of the most thoroughly researched neurological disorders within the realm of ML [50]. Techniques in supervised learning, including support vector machines (SVMs) and convolutional neural networks (CNNs), have been extensively utilized to distinguish between Alzheimer's patients and healthy individuals by analyzing neuroimaging data, such as MRI and PET scans [46]. Stumpo et al. [30] illustrated the effectiveness of CNNs in examining structural and functional brain imaging data for the identification of early biomarkers of AD. In a similar vein, Zhu et al. [6] employed ensemble methods, including random forests, to forecast early-stage AD by analyzing alterations in hippocampal volume, which serves as a crucial marker of disease advancement.

  The integration of ML with genomic data has significantly improved the predictive accuracy of AD models [44, 50]. According to Yi Huang et al. [4], multimodal approaches integrating imaging and genetic information demonstrate greater prediction accuracies than models relying on a single modality. Furthermore, federated learning frameworks have surfaced as an encouraging

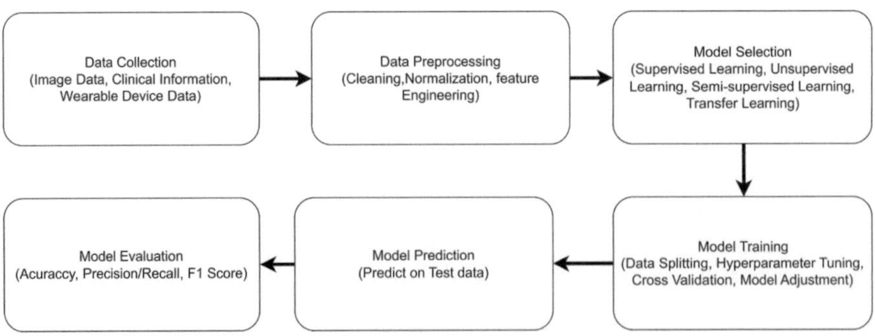

Fig. 2.8 Disease prediction workflow

approach to enhance model generalization while safeguarding data privacy, especially in multi-center studies that involve sensitive patient information [37, 38].

- **Case Studies on Parkinson's Disease**

  Significant promise has been shown by ML in the early detection and progression tracking of PD. Using gait analysis data, Yue Wang et al. [5] studied the use of ML models and effectively obtained great accuracy in separating PD patients from healthy controls. Li Shen and Thompson [32] investigated how recurrent neural networks (RNNs) may be used to analyze wearable sensor data, therefore enabling real-time tracking of motor symptom development. Furthermore, MacDonald et al. [34] demonstrated how well ensemble learning methods forecast patient responses to dopamine treatment—a necessary component of PD management.

  To improve diagnosis accuracy, recent developments in deep learning have made it possible to integrate multimodal data—including imaging, speech patterns, and handwriting analysis. Chen et al. [8] investigated how deep neural networks might be used to synthesize different data sources, therefore offering a better understanding of PD disease and its evolution.

- **Case Studies on Multiple Sclerosis (MS)**

  Clustering algorithms, a key aspect of unsupervised learning techniques, have played a crucial role in identifying disease subtypes and predicting progression in multiple sclerosis (MS). Smith et al. [35] employed ML models to categorize multiple sclerosis patients into progressive or relapsing-remitting types using MRI characteristics and clinical information. Zhu et al. [6] further illustrated the effectiveness of SVMs in predicting disease progression through the examination of alterations in lesion volume over time.

  Methods of Explainable AI (XAI) have garnered significant attention in the field of MS research. Clinicians are provided with an interpretable understanding of lesion development and therapy effects. These methods combine complex algorithms with useful medical knowledge, hence improving the clinical integration of ML models [37].

2.5.2 Imaging Analysis

Neuroimaging stands out as one of the most groundbreaking applications of machine learning (ML) in neurology [36, 39, 52]. Techniques like CNNs and generative models have revolutionized the analysis of MRI, CT, and PET scans, enabling more precise and efficient diagnoses. As depicted in Fig. 2.9, the image analysis workflow begins with Image Collection, which includes data from MRI, PET, CT, EEG, and optical imaging. This is followed by the Image Preprocessing stage, involving noise reduction, contrast enhancement, registration, segmentation, normalization, and augmentation. The next phase, Feature Extraction, employs methods such as histogram analysis, edge detection, texture analysis, shape analysis, autoencoders, and spatial frequency analysis. Subsequently, the Model Selection phase explores supervised, unsupervised, semi-supervised, and transfer learning approaches. The Model Training phase includes data splitting, hyperparameter tuning, cross-validation, and model adjustment. Finally, the Model Prediction phase applies the trained model to a test dataset, and the Model Evaluation phase assesses performance using metrics like accuracy, precision/recall, and F1 score. This comprehensive workflow underscores the transformative impact of ML in advancing neuroimaging and improving diagnostic accuracy in neurology.

In high-resolution neuroimaging data, ML is especially good in spotting important features. Stumpo et al. [30] demonstrated how CNNs are used for segmenting brain tumors and lesion detection in stroke patients, hence reducing the time and effort required for manual analysis. Zhu et al. [6] claim that using autoencoders has improved low-resolution scan quality, hence increasing diagnosis accuracy.

Graph-based neural networks (GNNs) have lately been used in the investigation of functional connectivity. Shen et al. [32] showcased the effectiveness of GNNs in identifying disruptions within brain networks linked to schizophrenia and autism. Matthew et al. [33] highlighted how ML plays a crucial role in automating manual segmentation tasks, like labeling brain regions in structural MRI scans, which have traditionally demanded significant expertise and time.

Neuroimaging research has benefited much from generative adversarial networks (GANs) and other deep generative models [49]. These models allow missing or corrupted scan sections to be synthesized, therefore extending the spectrum of research opportunities in data-limited settings. Furthermore, addressing data shortages and privacy issues, they help create synthetic datasets for the construction of strong ML models [35].

2.5.3 *Personalized Medicine*

Personalized medicine considers unique genetic, clinical, and lifestyle factors to help to customize therapies for individual patients. By aggregating multimodal data to predict treatment results and propose focused interventions, ML has become increasingly important in reaching this goal.

Huang et al. [4] underlined the application of ML for individual reactions to pharmaceutical treatments for epilepsy. By means of historical data and genetic

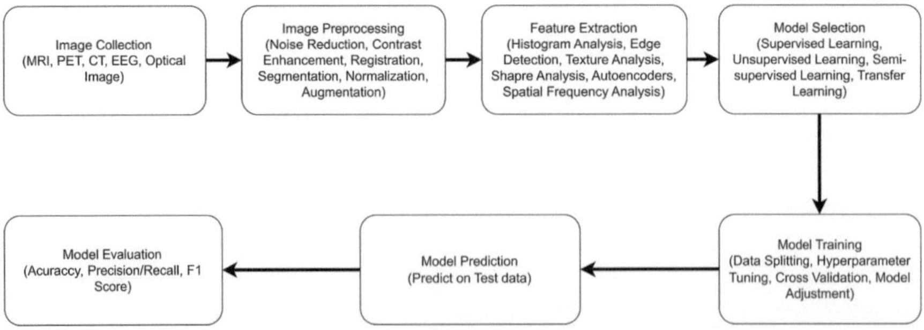

Fig. 2.9 Image analysis workflow

markers, ML models can find the most successful drugs while reducing adverse effects, hence improving patient outcomes. In a similar vein, MacDonald et al. [34] explored the application of RL to optimize deep brain stimulation (DBS) parameters for patients with PD, leading to improved symptom management and a better quality of life.

Furthermore, recently under investigation is the use of ML in stroke rehabilitation. Wang et al. [5] showed how ML models, trained on sensor data from rehabilitation equipment, can adjust therapy intensity in line with patient development, therefore ensuring individualized and successful recovery techniques. The advancements highlight the potential of ML to transform treatment strategies in neurology.

2.5.4 Monitoring and Management

Particularly in relation to chronic neurological diseases, ML has become increasingly important for tracking disease development and evaluating therapy efficacy. By combining real-time monitoring tools with longitudinal data, ML models give doctors practical insights meant to improve patient care. In Fig. 2.10, we can see the complete monitoring and management process. The flowchart illustrates a comprehensive approach to monitoring and managing chronic neurological disorders using advanced technologies and data-driven methodologies. It highlights the application of time series models and predictive analytics in Alzheimer's disease to track cognitive decline and facilitate proactive interventions, particularly during the transition from mild cognitive impairment to dementia. For multiple sclerosis, clustering algorithms are employed to monitor lesion volume changes, providing critical insights into disease activity and treatment response. Epilepsy management leverages machine learning-based seizure prediction and real-time analysis of EEG data, enhancing patient safety and quality of life. In Parkinson's disease, machine learning models trained on smartwatch data offer real-time feedback to clinicians, enabling more personalized and timely care. The integration of wearable technology and federated learning ensures privacy while facilitating remote patient monitoring, making healthcare more accessible and efficient. This holistic approach underscores the potential of technology in transforming the management of chronic neurological conditions.

ML models have been applied in AD to predict rates of cognitive deterioration, therefore enabling preventive therapies. Time-series models—including RNNs—were applied by Smith et al. [35] to examine longitudinal data from clinical trials, so effectively forecasting the changes from mild cognitive impairment to dementia. Monitoring changes in lesion volume for multiple sclerosis has shown the success of clustering techniques in offering the necessary understanding of disease activity and treatment response [6]. Likewise, real-time EEG data in ML-based seizure prediction systems for epilepsy generates preemptive warnings, improving patient safety and quality of life [31]. The combination of IoT devices and wearable technology has greatly expanded the function of ML in neurology. Using smartwatch data, Wang et al. [5] investigated the use of ML models to track tremors in PD

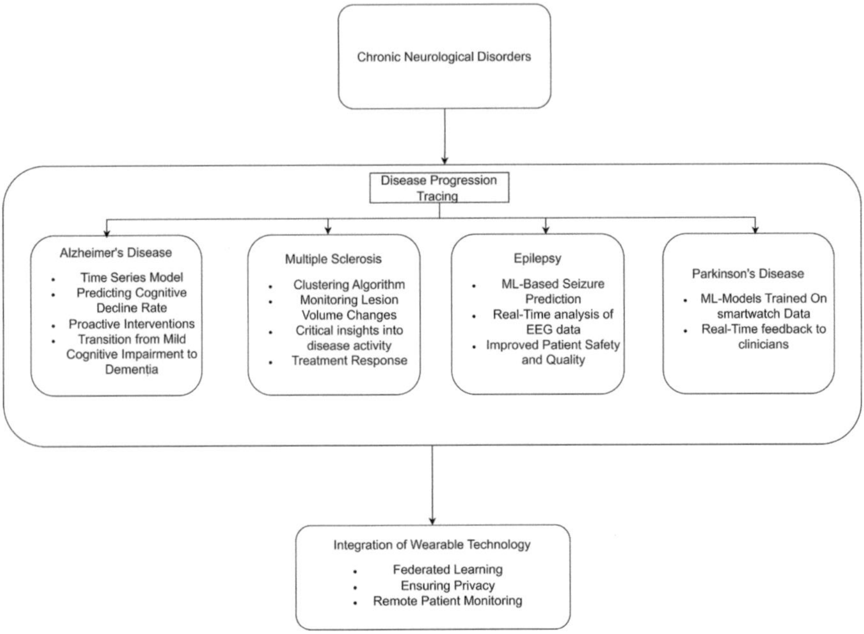

Fig. 2.10 Monitoring and management flowchart

patients, therefore providing instantaneous input to medical practitioners. These systems are increasingly including federated learning methods to improve data privacy and scalability in remote patient monitoring [38].

2.6 Evaluation Metrics

The development and application of ML models depend on their evaluation of them, particularly in healthcare environments where the effect of model errors can greatly influence patient outcomes [41, 42]. While underlining their relevance in neurology and other healthcare disciplines, this subsection explores the relevance of model evaluation by stressing common evaluation metrics and cross-validation approaches. Powerful assessment systems allow researchers and doctors to ensure that ML models are dependable, robust, and able to generalize successfully to fresh data [41].

2.6.1 Importance of Model Evaluation

An important stage in the development of ML systems, especially in the healthcare industry where the consequences are somewhat large, is evaluating models. These models directly affect human life by influencing choices about risk prediction, disease diagnosis, and treatment strategy. An efficient evaluation system ensures that models keep correctness, dependability, and generalizing capacity to new, unprocessed data. In neurology, this is especially important since the complex character and variety of neurological diseases call for models that can fit different patient groups and clinical settings.

Finding and correcting biases, verifying hypotheses, and improving performance all depend on accurate model evaluation. Biases in training data, for instance, can produce models that perform exceptionally well for some demographics while finding it difficult to extend to underrepresented groups. In healthcare, where guaranteeing equal access to accurate diagnosis and treatments is crucial, these prejudices can have major effects. Studies have repeatedly underlined the need of robust assessment systems in healthcare since they ensure the safety, efficacy, and equality of ML models and thereby guarantee their relevance [20, 24, 26].

Establishing confidence with stakeholders and doctors depends also on model evaluation. The proven dependability and interpretability of ML tools in healthcare usually define their application in this field. By means of extensive model evaluation, researchers can provide evidence-based insights on their performance, therefore improving the confidence in their therapeutic relevance.

2.6.2 Common Evaluation Metrics

Using a wide range of criteria, each of which offers a unique viewpoint on their strengths and shortcomings, the efficiency of ML models is assessed. In healthcare applications especially, where false positives and false negatives can have significant effects, these measures are particularly important [47]. Some of the most often used evaluation measures in neurology and allied disciplines are discussed below.

2.6.2.1 Accuracy

A basic statistic, accuracy measures the proportion of cases accurately predicted in respect to the overall count. Models are frequently assessed from this perspective, as it offers a comprehensive assessment of their capabilities. In Eq. 2.1, we can see that the ratio between the number of correct predictions and total number of predictions in percentage results in accuracy. In Eq. 2.2, it is shown using properties of the confusion metrics. Accuracy alone can be deceptive, particularly in datasets that are imbalanced, with one class having a substantially larger number of members than the other. For instance, in a dataset in which 95% of patients do not have a specific neurological disorder, a model that consistently predicts "no disorder" would accomplish 95% accuracy but would be clinically ineffective. Consequently, accuracy is a valuable initial metric; however, to conduct a thorough assessment, it is necessary to incorporate additional metrics [7, 20].

The accuracy equation is as follows:

$$\text{Accuracy} = \frac{\text{Number of Correct Predictions}}{\text{Total Number of Predictions}} \times 100\% \tag{2.1}$$

Mathematically,

$$\text{Accuracy} = \frac{TP + TN}{TP + TN + FP + FN} \tag{2.2}$$

where,

- TP = True Positives (Correctly predicted positive instances)
- TN = True Negatives (Correctly predicted negative instances)
- FP = False Positives (Incorrectly predicted positive instances)
- FN = False Negatives (Incorrectly predicted negative instances)

- ***Precision and Recall***

In healthcare applications, precision and recall are essential metrics, especially when false positives and false negatives have substantial repercussions. Often known as positive predictive value, in Eq. 2.3 we stated that precision is the ratio of actual positive predictions to the overall expected positives. In applications like cancer diagnosis, where false positives could lead to pointless and intrusive treatments, it is particularly important [46]. In Eq. 2.4, we can see that the ratio of true positives to the actual positives in the dataset is measured by recall, which is also referred to as sensitivity. In situations where the absence of a positive case (e.g., the failure to diagnose a neurological disorder) can result in grievous repercussions, it is essential [10, 22].

For instance, in the context of AD, a high recall guarantees that most patients with the condition are accurately identified [42]. Conversely, a high degree of precision guarantees that the individuals diagnosed will likely have the disease. It is frequently difficult to maintain a balance between precision and recall, as the enhancement of one often necessitates the sacrifice of the other [46].

$$\text{Precision} = \frac{TP}{TP + FP} \tag{2.3}$$

$$\text{Recall} = \frac{TP}{TP + FN} \tag{2.4}$$

- ***F1 Score***

The F1 score is a metric that balances precision and recall by taking the harmonic mean of the two [49]. The formula of F1 score is given in Eq. 2.5, and it's further clarified in Eq. 2.6. It is especially beneficial in healthcare scenarios that involve datasets that are imbalanced. For example, the F1 score provides a more complex assessment of model performance than accuracy alone in the diagnosis of rare neurological disorders [44]. The F1 score guarantees that models are assessed on their capacity to generate clinically pertinent predictions by considering both false positives and false negatives [7, 20, 45].

F1 Score equation:

$$F1\,\text{Score} = 2 \times \frac{\text{Precision} \times \text{Recall}}{\text{Precision} + \text{Recall}} \tag{2.5}$$

Substituting the formulas for Precision and Recall:

$$F1\,Score = 2 \times \frac{\frac{TP}{TP + FP} \times \frac{TP}{TP + FN}}{\frac{TP}{TP + FP} + \frac{TP}{TP + FN}} \tag{2.6}$$

- ***AUC-ROC (Area Under the Receiver Operating Characteristic Curve)***

The AUC-ROC (Area Under the Receiver Operating Characteristic Curve) metric is a widely used measure for evaluating the performance of binary classification models. It quantifies the model's ability to distinguish between the two classes by analyzing the trade-off between the true positive rate (TPR) and the false positive rate (FPR) across various classification thresholds. The TPR, also known as recall, is calculated as the ratio of true positives to the sum of true positives and false negatives (Eq. 2.7):

$$TPR = \frac{TP}{TP + FN} \tag{2.7}$$

The FPR, which represents the proportion of false positives relative to the sum of false positives and true negatives, is given by (Eq. 2.8):

$$FPR = \frac{FP}{FP + TN} \tag{2.8}$$

The ROC curve is constructed by plotting TPR against FPR at different threshold settings. The area under this curve (AUC) provides a single scalar value that summarizes the model's performance across all thresholds. Mathematically, the AUC is defined as the integral of the TPR with respect to the FPR (Eq. 2.9):

$$\text{AUC} = \int_0^1 \text{TPR}(\text{FPR})\,\text{d}(\text{FPR}) \tag{2.9}$$

In practice, it is approximated using the Trapezoidal Rule:

$$\text{AUC} = \sum_{i=1}^{n-1} \left(\text{FPR}_{i+1} - \text{FPR}_i\right) \times \frac{TPR_{i+1} + TPR_i}{2} \tag{2.10}$$

A model with an AUC-ROC of 1.0 indicates perfect discrimination between the classes, while an AUC-ROC of 0.5 suggests that the model performs no better than random guessing. This metric is particularly valuable in neurology for assessing models designed to predict the presence or absence of disorders, such as distinguishing between Alzheimer's patients and healthy individuals [22, 26].

- ***Specificity***

Specificity, also known as the true negative rate (TNR), measures a model's ability to correctly identify negative instances out of all actual negatives in the dataset. It is calculated as the ratio of true negatives (*TN*) to the sum of true negatives and false positives (*FP*) (Eq. 2.11):

$$\text{Specificity} = \frac{TN}{TN + FP} \tag{2.11}$$

Here, *TN* represents the number of correctly predicted negative instances, while FP denotes the number of incorrectly predicted positive instances. Specificity plays a critical role in healthcare settings, where false positives can have significant consequences. For example, in the diagnosis of multiple sclerosis, a high specificity ensures that patients who do not have the disease are not misdiagnosed, thereby avoiding unnecessary treatments and reducing patient anxiety.

When specificity is combined with recall (or true positive rate), it provides a comprehensive evaluation of a model's performance by ensuring that both positive and negative cases are accurately identified. This balance is essential for achieving a fair and reliable assessment of a model's effectiveness in real-world applications, particularly in medical diagnostics [24, 26].

2.6.3 *Methods for Cross-Validation*

Cross-valuation methods are important for assessing the generalizability and robustness of ML models. These techniques divide the data into several groups, iteratively train the model on certain subsets, and validate the model on others. This process ensures that the model is assessed on every data point, hence lowering the possibility of overfitting and so optimizing its potential to generalize to hitherto unheard-of data.

2.6.3.1 K-Fold Cross-Validation

K-fold cross-valuation is among the most often used techniques for model assessment. According to this approach, the dataset is split into K subsets (or folds), and the model is trained K times, each time using a separate fold as the validation set and the remaining folds as the training set [42]. Then, the average performance measures throughout the K iterations evaluate the model holistically. K-fold cross-valuation error has a generic formula provided by:

$$CV_{error} = \frac{1}{K}\sum_{i=1}^{K} E_i \tag{2.12}$$

where E_i represents the error on the ith fold, if accuracy is used as the evaluation metric, the formula is:

$$CV_{accuracy} = \frac{1}{K}\sum_{i=1}^{K} A_i \tag{2.13}$$

where A_i represents the accuracy on the ith fold. In the healthcare industry, where datasets are often limited and reliability is essential, this approach is very useful [6, 14, 24].

2.6.3.2 Stratified Cross-Validation

A variation on K-fold cross-valuation, stratified cross-valuation ensures that the class distribution in every fold corresponds with that of the whole dataset. Given sometimes skewed datasets in healthcare applications, this is extremely important. For PD, for example, the total number of patients in a dataset can be much less than the total number of healthy individuals. Stratified cross-validation ensures that every fold has a reasonable mix of both classes, therefore allowing a more precise evaluation of the performance of the model [20, 21, 25].

2.6.3.3 Leave-One-Out Cross-Validation (LOOCV)

Leave-one-out cross-validation is an extreme variation of K-fold cross-validation, in which K is the number of samples in the dataset. The model is trained on the remaining samples, and a single sample is used as the validation set in each iteration. Small datasets benefit especially from LOOCV as, despite its processing cost, it provides an objective estimate of model performance. Given patient data may be limited in neurology, LOOCV might offer insightful analysis on the generalizability of models [21, 25].

2.6.3.4 Time-Series Cross-Validation

Applications using longitudinal data, such as the tracking of disease development in Alzheimer's or Parkinson's illness, routinely use time-series cross-validation. This method guarantees that the model is validated on future time points about the training data. During the simulation of real-world scenarios, the model is required to forecast outcomes using historical data. Maintaining the temporal structure of the data, time-series cross-valuation provides a more realistic evaluation of the predictive powers of the model [21, 25].

2.7 Ethical Considerations

Particularly in neurology, the integration of ML into healthcare raises several ethical questions that must be answered if we are to ensure the appropriate and fair use of these technologies. Given the increasing impact of ML models on clinical decision-making, it is imperative to address issues related to data privacy, bias, impartiality, and the function of doctors in the deployment and supervision of these systems [41]. Still, these technologies also have the power to greatly improve clinical judgment. Emphasizing the challenges and necessary solutions to ensure that ML applications in healthcare are both morally sound and successful, this subsection explores these ethical aspects.

2.7.1 Data Privacy Concerns

Implementing ML in the healthcare industry depends on the collection and analysis of large volumes of sensitive patient data, which raises significant questions about data privacy and security [44]. With patient confidentiality as a basic principle of medical ethics, the management of personal health information (PHI) must follow strict regulatory frameworks, including the General Data Protection Regulation (GDPR) in Europe and the Health Insurance Portability and Accountability Act (HIPAA) in the United States. These rules demand the application of thorough procedures to protect patient data, including anonymization, encryption, and access limits, thereby preventing illegal access and data breaches [6, 17, 24, 41].

To protect patient privacy, it is imperative to implement anonymization strategies that eliminate or obscure personally identifiable information (PII). Nevertheless, anonymization alone is not always sufficient, as new data re-identification techniques have demonstrated the potential to reverse engineer anonymized data.

Encryption provides an additional layer of security that ensures that data is protected even if it is intercepted or accessed without authorization, both in transit and at rest. Data exposure is further restricted to only those individuals who require it for legitimate purposes through access controls, such as role-based permissions and audit traces.

Combining the need to safeguard patient privacy with the requirement for data availability for model training is somewhat difficult. Common practice in conventional methods is centralizing data in a single repository, which limits data sharing among organizations and raises the possibility of breaches. Federated learning and differential privacy are emerging viable alternatives. Federated learning facilitates the training of models across decentralized datasets without transmitting raw data, thereby conserving privacy and facilitating collaborative research [43]. Differential privacy, on the other hand, uses mathematical noise in datasets to avoid the identification of individual patients while nevertheless enabling the meaningful analysis to be performed [22, 25, 26].

2.7.2 *Bias and Fairness in ML Models*

In ML models, bias raises serious ethical questions, especially in the healthcare industry, as skewed forecasts could lead to unfair results and aggravation of already existing inequalities. Systemic inequalities in healthcare access and delivery, defective algorithms, and skewed training data are all potential sources of bias. For instance, the datasets utilized to train ML models may under-represent specific demographic groups, such as racial minorities or individuals from low-income backgrounds, resulting in models that perform inadequately for these populations [8, 13, 24].

To handle bias, a multifarious strategy starting with careful data preparation is required. Synthetic data generation, reweighting, and resampling are among the techniques one can use to guarantee models are trained on more fair datasets, therefore balancing representation among underrepresented groups. Whereas reweighting emphasizes samples that are under-represented throughout the model training phase, resampling is the adjustment of the distribution of the dataset to assure that all groups are sufficiently represented. Using generative adversarial networks (GANs), synthetic data generation can create extra training instances for underrepresented groups, therefore enhancing the equity of the model [10, 21].

Moreover, essential are algorithmic transparency and interpretability if we are to find and reduce biases in ML models. By means of insights into the decision-making processes of models [42], XAI techniques—including SHAP (SHapley Additive exPlanations) and LIME (Local Interpretable Model-agnostic Explanations)—offer physicians and researchers the means to identify and correct biased outcomes. Moreover, including fairness restrictions into model development helps to ensure that forecasts are fair for many demographic groups. Fairness-aware algorithms can be created, for instance, to minimize the differences in false positive and false negative rates between groups, therefore promoting more fair healthcare results [22, 26, 41].

For a given feature x_i, its SHAP value is calculated as:

$$\phi_i = \sum_{S \subseteq F \setminus \{i\}} \frac{|s|!\left(|F|-|S|-1\right)!}{|F|!}\left[f\left(S \cup \{i\}\right) - f\left(S\right)\right] \tag{2.14}$$

where:

- F is the set of all features.
- S is a subset of F, **excluding** x_i.
- $f(S)$ is the model's prediction using only the features in subset S.
- ϕ_i is the contribution (SHAP value) of feature x_i.

LIME fits a **local linear model** $g(x)$ around a given instance x by solving:

$$\arg\min_g \sum_i \pi_x(z_i)\left(f(z_i) - g(z_i)\right)^2 + \Omega(g) \tag{2.15}$$

where:

- $f(z_i)$ is the black-box model's prediction.
- $g(z_i)$ is the interpretable surrogate model's prediction.
- $\pi x(z_i)$ is a kernel function giving weight to instances close to x.
- $\Omega(g)$ is a complexity penalty for g.

2.7.3 The Role of Clinicians

In healthcare applications, human supervision is still vital even if ML models have growing power. The validation of model suggestions and the contextualization of predictions within the larger framework of patient care depend much on the actions of clinicians. Their knowledge is priceless in making sure that the insights given by ML models fit patient-centered care approaches and are clinically meaningful. Although ML models can offer valuable insights and facilitate decision-making [54], they should not supplant the clinical judgment and expertise of healthcare professionals. Rather than operating autonomously, ML systems should serve as decision-support tools, enhancing the capabilities of clinicians [7, 10, 24].

Working together, data scientists and doctors can ensure that ML models are not only clinically relevant and consistent with patient-centered care procedures but also technically accurate [41]. Interpretation of model findings and the development of well-informed judgments depend on this multidisciplinary cooperation. Clinicians are extremely important in this procedure because of their subject knowledge and sharp awareness of patient care. On the other hand, data scientists provide technical expertise in the development and optimization of ML models. Healthcare institutions can create ML systems that are both ethically sound and effective by encouraging and fostering this interdisciplinary collaboration, thereby ensuring that all contributions are valued and utilized to their best advantage [21, 25, 41].

The encouragement of clinician education and training on AI technologies is another vital element of the effective integration of ML into healthcare systems. Many doctors not knowing the basic ideas of ML could lead to skepticism of AI-driven suggestions. By allowing doctors to confidently use ML technologies in their daily work and using resources and training programs, one can help to close this knowledge gap. Involving ML models in designing and assessing these technologies helps doctors to build their trust and acceptance of them, therefore guaranteeing their best use [6, 13, 22].

2.8 Future Directions

The fast development of ML in neurology has revealed new possibilities in the diagnosis, treatment, and understanding of neurological diseases. As the field develops, several emergent themes and issues are influencing its course and present chances as well as difficulties for their incorporation into clinical practice. This segment explores these developments, with special focus on the difficulties of including these technologies into clinical processes and the developing trends in ML for neurology. The ethical issues defined in the section before are discussed in more detail, underlining the need for flexibility, openness, and teamwork in the application of ML systems.

2.8.1 Emerging Trends in ML for Neurology

Several cutting-edge advancements are revolutionizing the application of ML in neurology. Each has the potential to broaden the scope of ML-driven solutions and resolve enduring limitations [44]. Federated learning, multimodal data integration, and XAI are particularly promising trends.

- ***Addressing the "Black-Box" Issue: Explainable AI (XAI)***

One of the main reasons ML is not widely adopted in neurology is the intrinsic opacity of many ML models, sometimes referred to as the "black box" problem. Clinicians find it difficult to understand the process by which predictions or recommendations are produced as standard ML models—especially deep learning architectures—are often criticized for their interpretability [41]. In neurology, therapy and diagnostic choices often have major effects on patient outcomes; hence, lack of openness might affect trust and approval among medical practitioners.

XAI has emerged as a critical research area designed to address this challenge. XAI concentrates on the development of methodologies and techniques that enhance the transparency and interpretability of the decision-making processes of ML models. Insights into the factors that influence model predictions are provided by XAI, which allows clinicians to assess the reliability and validity of ML-driven recommendations. For example, in the field of neuroimaging, XAI techniques such as saliency maps or attention mechanisms can emphasize the precise regions of an image that are most important in determining a diagnosis, thereby providing clinicians with a more comprehensive comprehension of the model's reasoning [13, 26, 28].

The significance of XAI in neurology is immeasurable. Complex and multifactorial etiologies are frequently associated with neurological conditions, necessitating a nuanced comprehension of the underlying mechanisms. XAI fosters transparency, which in turn enhances clinician trust and enables the identification of potential biases or errors in model predictions. This is especially important in high-stakes situations, such as diagnosing neurodegenerative diseases or predicting treatment responses, where the repercussions of erroneous decisions can be severe [41].

- ***Federated Learning: Improving Collaboration and Data Privacy***

The adoption of federated learning, a decentralized approach to model training that addresses the challenges of data privacy and accessibility [43], is another transformative trend in ML for neurology. In the context of sensitive neurological data, traditional ML models typically necessitate centralized datasets, which can pose significant risks to patient confidentiality and raise regulatory concerns. These issues are resolved through federated learning, which facilitates collaborative model training across multiple institutions without the necessity of sharing raw data. Rather, the model is trained locally on the data of each participating institution, and only the model updates (e.g., gradients) are shared and aggregated to enhance the global model [17, 21, 26, 43].

There are numerous benefits to this method in the field of neurology. Initially, it guarantees adherence to stringent data privacy regulations, including the General Data Protection Regulation (GDPR) and the Health Insurance Portability and Accountability Act (HIPAA), by maintaining patient data in a localized location. Secondly, federated learning enables the consolidation of a variety of datasets from geographically dispersed sources, thereby improving the generalizability of ML models [54]. This is especially advantageous in neurology, where uncommon conditions or small sample sizes may restrict the efficacy of models trained on single-institution datasets. For instance, the utilization of data from numerous specialized centers worldwide could facilitate the development of robust models for rare neurological disorders through federated learning.

- ***Multimodal Data Integration: Broadening the Application of Machine Learning Models***

Another substantial advancement in neurology's application of ML is the incorporation of multimodal data. A comprehensive approach to diagnosis and treatment is often required for neurological disorders, which are frequently defined by intricate interactions between genetic, environmental, and clinical factors. Traditional ML models frequently depend on solitary data modalities, such as imaging or clinical records, which may only offer a partial understanding of the underlying pathology. By integrating a variety of data sources, such as neuroimaging, genetic information, electrophysiological data, and clinical histories, into a unified analytical framework, multimodal data integration aims to address this limitation [10, 20].

This method has the potential to substantially improve the accuracy and scope of ML models in neurology. For example, combining MRI scans with genetic markers and patient-reported outcomes could enhance the early detection of neurodegenerative diseases, such as Alzheimer's or Parkinson's. Similarly, integrating clinical assessments with EEG data could improve the diagnosis and monitoring of epilepsy. Multimodal data integration facilitates more personalized and comprehensive patient care by documenting the multifaceted character of neurological disorders.

2.8.2 The Integration of Machine Learning into Clinical Practice

Despite their immense potential for neurology, successful assimilation of ML advancements into clinical practice remains a formidable challenge. These technologies must be realized by overcoming various obstacles, such as regulatory impediments, infrastructure limitations, and clinician resistance.

- ***Standardized Frameworks for Model Deployment***

The absence of standardized frameworks for model deployment is one of the primary impediments to the integration of ML into clinical practice. The seamless integration of ML systems with existing clinical protocols [53], such as electronic health records (EHRs), diagnostic tools, and decision-support systems, is frequently necessary for the deployment of ML systems in healthcare settings [41]. Nevertheless, the absence of standardized protocols and interfaces complicates this process, resulting in potential disruptions and inefficiencies in clinical practice [6, 13, 24, 46].

Researchers and industry stakeholders are increasingly advocating for the development of adaptable and modular ML systems that can be readily integrated into diverse healthcare environments to address this challenge [45]. These systems should be developed with interoperability in mind, guaranteeing compatibility with clinical tools and EHR platforms that are frequently employed. Furthermore, the development of open-source frameworks and guidelines for model deployment could facilitate the adoption of ML technologies across institutions, thereby alleviating the burden on individual healthcare providers.

- ***Establishing Clinician Trust and Competence***

Another critical factor in the successful integration of ML systems is the establishment of clinician trust. Numerous healthcare providers continue to harbor reservations regarding ML-driven instruments, frequently because of their inadequate comprehension of the fundamental principles or their reservations regarding the accuracy of model predictions. A multifaceted approach that integrates education, transparency, and collaboration is necessary to address this skepticism.

Educational initiatives designed to equip clinicians with a fundamental comprehension of ML principles and practical experience with AI tools can be instrumental in developing trust and competence. Empowering clinicians to critically evaluate their outputs and demystifying ML technologies can be achieved through training programs, seminars, and online courses. Furthermore, by encouraging interdisciplinary collaboration between clinicians and data scientists, the distance between technical expertise and clinical knowledge can be bridged, guaranteeing that ML systems are designed with end-users' requirements and perspectives in mind [7, 21].

- ***Decision-Support Tools and Hybrid Models***

Another approach to incorporating ML into clinical practice is the use of hybrid models in which ML systems function as decision-support instruments rather than autonomous decision-makers [53]. In this framework, clinicians can override or disregard recommendations when necessary while utilizing ML-driven insights to inform their decisions. They retain the ultimate authority over patient care. This method alleviates apprehensions regarding the dependability of ML systems and is consistent with the ethical requirement to prioritize clinician judgment and patient autonomy [24, 26].

- ***Ethical and Regulatory Considerations***

Careful attention to regulatory and ethical considerations is also required when integrating ML into clinical practice. The safety and efficacy of ML-driven medical devices are contingent upon the acquisition of regulatory sanction, such as authorization from the U.S. Food and Drug Administration (FDA) [41]. Nevertheless, the regulatory frameworks that were developed for inert technologies face distinct challenges due to the dynamic nature of ML models, which frequently necessitate continuous updates and retraining [55]. The development of adaptive regulatory pathways that accommodate the iterative character of ML systems will be necessary to address these challenges [20, 21, 26, 55].

- ***Feedback Loops and Iterative Development***

Lastly, the successful incorporation of ML into clinical practice necessitates an iterative approach that integrates continuous feedback from patients and clinicians. By establishing feedback channels between healthcare providers and data scientists, ML systems can be continuously refined to ensure clinical relevance, usability, and adaptability. This collaborative process not only improves the performance of ML models but also cultivates a sense of ownership and engagement among clinicians, thereby fostering their acceptance and adoption [20, 21, 26].

2.9 Conclusion

The use of ML in neurology has demonstrated significant potential in enhancing the diagnosis, treatment, and management of neurological disorders, with models achieving remarkable accuracy, such as 96.3% in detecting conditions like Parkinson's disease. By integrating multimodal data sources, ML enables more precise disease characterization and personalized treatment approaches. In clinical settings, applications like ChatGPT-4 have shown high accuracy in providing epilepsy-related information, showcasing the practical benefits of ML in real-world healthcare scenarios. However, several challenges remain, including data security, ethical concerns, regulatory frameworks, and significant disparities in healthcare access, as over 80% of neurological deaths occur in low- and middle-income countries. These issues highlight the need for equitable and secure implementation of ML technologies. Looking ahead, the future direction of ML in neurology includes advancements in personalized medicine, sophisticated imaging analysis techniques, and the integration of diverse data modalities to achieve a comprehensive understanding of neurological diseases. To harness the potential of ML fully, future efforts must focus on developing standardized validation protocols, addressing data quality and accessibility issues, improving model interpretability, and fostering interdisciplinary collaboration between clinicians and ML experts. The integration of ML technologies with clinical expertise holds great promise for improving patient outcomes and advancing the understanding of neurological diseases. Future research should prioritize addressing existing limitations, ensuring ethical and interpretable ML systems, and exploring innovative approaches to adapt to the evolving healthcare landscape. Key future directions include personalized treatment strategies, advanced imaging and data analysis, and the development of robust, equitable, and collaborative frameworks for ML in neurology.

Future Directions

- ***Explainable AI (XAI):*** Development of transparent models to enhance clinician trust and regulatory compliance.
- ***Federated Learning:*** Decentralized training to improve data privacy and model generalizability.
- ***Multimodal Integration:*** Combining neuroimaging, genomic, and clinical data for holistic disease understanding.
- ***Regulatory Adaptation:*** Dynamic frameworks to accommodate evolving ML systems in clinical practice.

By concentrating on these important areas, the discipline may effectively employ ML in neurology, ultimately transforming the way neurological care is delivered and improving outcomes for patients worldwide.

References

1. W. H. Organization. Over 1 in 3 people affected by neurological conditions, the leading cause of illness and disability worldwide. World Health Organization; 2024.
2. Owolabi MO, et al. The neurology revolution. The Lancet Neurology. 2022;21(11):960–1.
3. Gourie-Devi M. Epidemiology of neurological disorders in India: review of background, prevalence and incidence of epilepsy, stroke, Parkinson's disease and tremors. Neurol India. 2014;62:588.
4. Huang Y, Li Y, Pan H, Han L. Global, regional, and national burden of neurological disorders in 204 countries and territories worldwide. J Glob Health. 2023;13:04160.
5. Wang Y, et al. Burden of common neurologic diseases in Asian countries, 1990–2019: an analysis for the global burden of disease study 2019. Neurology. 2023;100(21):e2141–54.
6. Zhu Y, et al. Multimodal neuroimaging-based prediction of Parkinson's disease with mild cognitive impairment using machine learning technique. NPJ Parkinson's Disease. 2024;10(1):218.
7. Sunil G, et al. Graph neural network and machine learning analysis of functional neuroimaging for understanding schizophrenia. BMC Neurosci. 2024;25(1):2.
8. Chen K-C, et al. CompressedMediQ: hybrid quantum machine learning pipeline for high-dimensional neuroimaging data. arXiv preprint arXiv. 2024:2409.08584.
9. Huang Y, et al. Innovative neuroimaging biomarker distinction of major depressive disorder and bipolar disorder through structural connectome analysis and machine learning models. Diagnostics. 2024;14(4):389.
10. Lock C, et al. Neuroimaging data repositories and AI-driven healthcare—global aspirations vs. ethical considerations in machine learning models of neurological disease. Front Artifi Intell. 2024;6:1286266.
11. Galil A, Galil WA. Applications of artificial intelligence in the field of diagnostic medicine and future prospects. Int J Artifi Intell Emerg Technol. 2023;6(2):1–18.
12. R. Mukherjee, et al., An intelligent CDS (clinical decision support) framework using machine learning algorithms for Parkinson disease detection. Preprint, 2024.
13. Owusu G, Wang X, Sun J. Longitudinal healthcare analytics for early detection and progression of neurological diseases: a clinical decision support system. In: Proceedings of the 57th Hawaii international conference on system sciences (HICSS). Hawaii; 2024.
14. Masood A, et al. Review on enhancing clinical decision support system using machine learning. In: CAAI transactions on intelligence technology; 2024.
15. Tang B-H, et al. Optimal use of β-lactams in neonates: machine learning-based clinical decision support system. EBioMedicine. 2024;105
16. Wang L, Long D-Y. Significant risk factors for intensive care unit-acquired weakness: a processing strategy based on repeated machine learning. World J Clin Cases. 2024;12(7):1235.
17. Lefkovitz I, et al. Direct clinical applications of natural language processing in common neurological disorders: scoping review. JMIR Neurotechnol. 2024;3(1):e51822.
18. Kim H-W, et al. "assessing the performance of ChatGPT's responses to questions related to epilepsy: a cross-sectional study on natural language processing and medical information retrieval," *seizure: European*. J Epilepsy. 2024;114:1–8.
19. Lo Barco T, et al. Natural history of rare diseases using natural language processing of narrative unstructured electronic health records: the example of Dravet syndrome. Epilepsia. 2024;65(2):350–61.
20. Chiang C-C, et al. A large language model–based generative natural language processing framework fine-tuned on clinical notes accurately extracts headache frequency from electronic health records. Headache: J Head Face Pain. 2024;64(4):400–9.
21. Pedro KM, et al. Machine learning-based cluster analysis identifies four unique phenotypes of patients with degenerative cervical myelopathy with distinct clinical profiles and long-term functional and neurological outcomes. EBioMedicine. 2024;106

22. Husnain A, et al. A precision health initiative for chronic conditions: design and cohort study utilizing wearable technology, machine learning, and deep learning. Int J Adv Eng Technol Innov. 2024;1(2):118–39.
23. Yousefi M, et al. Machine learning based algorithms for virtual early detection and screening of neurodegenerative and neurocognitive disorders: a systematic-review. Front Neurol. 2024;15:1413071.
24. Terranova N, Venkatakrishnan K. Machine learning in modeling disease trajectory and treatment outcomes: an emerging enabler for model-informed precision medicine. Clin Pharmacol Therap. 2024;115(4):720–6.
25. Catacutan DB, et al. Machine learning in preclinical drug discovery. Nat Chem Biol. 2024;20(8):960–73.
26. Kalani M, Anjankar A. Revolutionizing neurology: the role of artificial intelligence in advancing diagnosis and treatment. Cureus. 2024;16(6):e61706.
27. Vicidomini C, Fontanella F, D'Alessandro T, Roviello GN. A survey on computational methods in drug discovery for neurodegenerative diseases. Biomolecul. 2024;14(10):1330.
28. Wei H, et al. Explainable machine learning for predicting neurological outcome in hemorrhagic and ischemic stroke patients in critical care. Front Neurol. 2024;15:1385013.
29. Miller MI, Shih LC, Kolachalama VB. Machine learning in clinical trials: a primer with applications to neurology. Neurotherapeutics. 2023;20(4):1066–80.
30. Stumpo V, et al. Machine learning algorithms in neuroimaging: an overview. In: Machine learning in clinical neuroscience: foundations and applications; 2022. p. 125–38.
31. Davatzikos C. Machine learning in neuroimaging: Progress and challenges. NeuroImage. 2019;197:652–6.
32. Shen L, Thompson PM. Brain imaging genomics: integrated analysis and machine learning. Proc IEEE. 2019;108(1):125–62.
33. Thomas RM, et al. Dealing with missing data, small sample sizes, and heterogeneity in machine learning studies of brain disorders. In: Machine learning. Academic Press; 2020. p. 249–66.
34. MacDonald BK, Cockerell OC, Sander JWAS, Shorvon SD. The incidence and lifetime prevalence of neurological disorders in a prospective community-based study in the UK. Brain. 2000;123(4):665–76.
35. Smith CM, Weathers AL, Lewis SL. An overview of clinical machine learning applications in neurology. J Neurol Sci. 2023;455:122799.
36. Castiglioni I, et al. AI applications to medical images: from machine learning to deep learning. Phys Med. 2021;83:9–24.
37. Jovel J, Greiner R. An introduction to machine learning approaches for biomedical research. Front Med. 2021;8:771607.
38. Albahadily HK, Mohammed MQ, Mohammed AK. A practical guide of machine learning algorithms and applications. Int J Appl Info Sys. 2023;12(40):8–13.
39. Dewan MC, et al. Estimating the global incidence of traumatic brain injury. J Neurosurg. 2018;130(4):1080–97.
40. English M, et al. Machine learning in neuro-oncology, epilepsy, Alzheimer's disease, and schizophrenia. In: Machine learning in clinical neuroscience: foundations and applications; 2022. p. 349–61.
41. Prasad A. Introduction to data governance for machine learning systems. Apress; 2024.
42. González-Rodríguez VE, Izquierdo-Bueno I, Cantoral JM, Carbú M, Garrido C. Artificial intelligence: a promising tool for application in phytopathology. Horticulturae. 2024;10(3):197.
43. Elngar A, Oliva D, Balas V. Artificial intelligence using federated learning: fundamentals, challenges, and applications. 1st ed. CRC Press; 2024.
44. Goyal D, Pratap B, Gupta S, Raj S, Agrawal R, Kishor I. Recent advances in sciences, engineering, Information Technology & Management. 1st ed. CRC Press; 2025.
45. Elhanashi A, Dini P. Machine learning for cybersecurity: threat detection and mitigation. MDPI; 2024.

46. Jena B, Saxena S, Paul S. Machine learning for neurodegenerative disorders: advancements and applications. CRC Press; 2025.
47. Bhatti U, Mengxing H, Li J, Bazai S, Aamir M. Deep learning for multimedia processing applications. In: Volume two: signal processing and pattern recognition. CRC Press; 2024.
48. Gururaj HL, Flammini F, Ravi Kumar V, Prema NS, Recent trends in healthcare innovation: proceedings of the Annual International Conference on Recent Trends in Healthcare Innovation (AICRTHI 2024)., Mysuru, India, October 24th–25th, 2024, CRC Press, 2025.
49. Bacon E, He D, Achi N. Neuroimage analysis using artificial intelligence approaches: a systematic review. Med Biol Eng Comput. 2024;62:2599–627.
50. Dominguez-Gortaire J, Ruiz A, Porto-Pazos A, Rodriguez-Yanez S, Cedron F. Alzheimer's disease: exploring pathophysiological hypotheses and the role of machine learning in drug discovery. Int J Mol Sci. 2025;26(3):1004.
51. Moiniafshari K, Zanut A, Tapparo A, Pastore P, Bogialli S, Monikh F. A perspective on the potential impact of microplastics and Nanoplastics on the human central nervous system. Environ Sci: Nano; 2025.
52. Liang M, Li P, Xie S, Huang X, Li X. Integrative prognostic modeling for stage III lung adenosquamous carcinoma post-tumor resection: machine learning insights and web-based implementation. Front Surg. 2024;11:1489040.
53. Busch F, Hoffmann L, Rueger C, van Dijk E, Kader R, Ortiz-Prado E. Systematic review of large language models for patient care: current applications and challenges. MedRxiv. 2024:2024–03.
54. Balaji D, Kalayci P, Ramkumar S. Advances in sports science and technology. CRC Press; 2025.
55. Ain QU, Nazir R, Nawaz A, Shahbaz H, Dilshad A, Mufti I. Machine learning approach towards quality assurance, challenges and possible strategies in laboratory medicine. J Clin Trans Pathol. 2024;4(2):76–87.

Chapter 3
Brain Connectomics and Graph Theory

Md. Mehedi Hassan

3.1 Introduction

Brain connectomics is the study of the complex brain connection networks, or the "connectome." The brain's connectome is an enormous and complicated web composed of structural and functional links connecting different parts of the brain. The connections allow the brain to process information, regulate body functions, and mediate cognitive and emotional reactions [1, 2].

There are two main fields of brain connectomics:

1. **Structural Connectomics**: The physical, anatomical connection between brain regions. It is the map of white matter pathways, composed of axons interconnecting neurons. These are measurable by employing neuroimaging techniques such as diffusion tensor imaging (DTI), an MRI scan that captures the direction and integrity of white matter tracts [3].
2. **Functional Connectomics**: This is the dynamic interactions between different brain regions that collaborate to execute a range of cognitive and behavioral operations. Functional connections are typically investigated with functional MRI (fMRI), which detects brain activity by measuring changes in levels of blood oxygenation associated with neuronal activity. Functional connectivity is the temporal correlations between brain regions that show co-activation during tasks or at rest [4].

Understanding the brain's structural and functional connectivity is key to demystifying normal brain function mechanisms and neurological and psychiatric diseases [5]. As an example, in disorders such as Alzheimer's disease or schizophrenia, these linkages are constantly disturbed or rewired and lead to cognitive and behavioral impairments.

M. M. Hassan (✉)
Computer Science and Engineering, Khulna University, Khulna, Bangladesh

M. M. Hassan et al. (eds.), *Graph Neural Networks for Neurological Disorders*,
https://doi.org/10.1007/978-3-032-04315-3_3

3.1.1 Graph Theory in Brain Connectomics

Graph theory gives us the mathematical and computational tools to analyze complex networks such as the ones in connectomics. In graph theory, the graph is characterized by nodes (vertices) and edges (connecting nodes) [5]. In brain connectomics:

- **Nodes**: Correspond to brain areas or regions of interest (ROIs), such as particular functional or anatomical regions of the brain.
- **Edges**: Describe the connectivity (either structural or functional) of these regions.

Brain networks can be represented as graphs with nodes corresponding to distinct brain regions (e.g., motor cortex, prefrontal cortex, visual cortex) and links representing the connections between the regions (e.g., axonal fibers in structural networks or temporal correlations in functional networks) [6].

3.1.2 How Graph Theory Works in Brain Connectomics

Graph theory allows researchers to quantify and analyze the brain's network attributes in an exact and mathematical way. From the study of network structure and organization within the brain, we are able to shed some light on significant properties of the connectivity of the brain beneficial not just for comprehension of normal cognition but also for research into disease processes [7]. Some of the most significant ideas from graph theory in brain connectomics are:

1. Global Network Properties

The brain network is of small-world nature, whereby most regions are no more than a few away from each other, but a few highly connected hubs (nodes) exist. This structure allows the brain to have efficient communication between regions without having a relatively expensive network. Brain networks are likely to be organized into modules or communities. These modules are collections of brain regions that are more connected to each other than to the rest of the brain. Functional modules may be equivalent to networks for specific cognitive abilities, such as the default mode network or the salience network. This refers to the rate and quality with which information can spread through the network. Efficient networks are necessary for quick decision-making and mental processing.

2. Local Network Properties

This measure defines the number of connections that a particular brain region has with other regions. A region with high centrality is referred to as a "hub" and is responsible for coordinating brain activity. This measure identifies brain regions that lie on many of the shortest paths between other regions. They are bridges, conveying information among widely separated brain regions. This measures how strongly the neighbors of a given node are connected to each other. High clustering indicates that an area is part of a tightly connected set of brain areas.

3.2 Combining Brain Connectomics with Graph Theory

The integration of brain connectomics and graph theory provides a more holistic view of brain function. By representing the brain's connectome as a graph, graph theory offers a powerful toolkit to analyze the structure and dynamics of these networks, uncovering insights about brain organization, cognition, and dysfunction.

3.2.1 Structural Brain Networks

Graph theory is a valuable tool in the field of structural connectomics to investigate the white matter networks of the brain. Formally, the brain may be modeled as a network where local areas (nodes) are connected by white-matter pathways (edges) [8]. Through the use of graph theory, scientists can spot these key hubs in the brain that serve as central conduits of communication between various areas—for example, the thalamus, or the default mode network. Also, this can be used to measure network efficiency, shedding light on the connectivity of different brain regions, and how efficiently information can travel through the brain.

3.2.2 Functional Brain Networks

Functional connectomics uses graph theory to study how different parts of the brain interact over time, whether you're doing a task or resting. Using fMRI, which is a type of brain scan, researchers can see which brain areas activate together. Graph theory helps in discovering important insights about the brain. It identifies groups of brain areas that collaborate on similar tasks, like moving or making decisions [9]. It also helps examine the brain's strength and flexibility, showing how it recovers from changes, injury, or disease. Another key aspect is finding important brain areas, known as hubs, that support complex tasks like focusing, remembering, and processing emotions. This shows how complex and active the connections in the brain really are.

3.3 Applications and Implications of Their Synergy

Using brain connectomics and graph theory together is very useful for understanding how the brain works, whether healthy or not. These tools are important in several areas:

1. Neurological and Psychiatric Disorders: Graph theory is used to spot unusual patterns in brain networks for diseases like Alzheimer's, schizophrenia, depression, and autism. For example, issues in specific brain areas might help explain thinking problems in these illnesses [10].
2. Cognitive Neuroscience: Scientists study how various brain parts cooperate to handle complex things like attention, memory, language, and decision-making. Graph theory shows which networks are strong and efficient for these mental tasks [11].
3. Brain Development and Aging: The network properties of the brain change as people grow older. By studying these changes through connectomics and graph theory, researchers can learn how aging affects brain connections and how early-life problems might impact mental growth in the long term.

3.4 Conclusion

The combination of brain connectomics and graph theory offers an effective approach to understanding the brain's structure and function. Brain connectomics is like a detailed map, showing how different areas of the brain connect. On the other hand, graph theory provides methods to analyze and interpret these connections using numbers and patterns. When connectomics and graph theory are used together, they help us learn a lot about how the brain organizes and processes information. This approach also helps us understand how problems in these brain networks might cause diseases. Additionally, it can assist in finding ways to improve brain function by targeting these networks. As technology in brain imaging and computing gets better, using connectomics and graph theory will become crucial in exploring the brain's complex nature.

References

1. Hassett JD, et al. Development of the whole-brain functional connectome explored via graph theory analysis. Aperture Neuro. 2024;4
2. Erciyes K. Complex brain networks: a graph-theoretical analysis. Bioinfo Brain. 2024:224–49.
3. Bhattacharya D, et al. Analyzing brain tumor Connectomics using graphs and persistent homology. In: International workshop on topology-and graph-informed imaging informatics. Cham: Springer; 2024.

4. Mei Y, et al. Disrupted topologic efficiency of white matter structural connectome in migraine: a graph-based connectomics study. J Headache Pain. 2024;25(1):204.
5. Mithani K, Suresh H, Ibrahim GM. Graph theory and modeling of network topology in clinical neurosurgery. Comput Neurosurg. 2024:107–22.
6. Chen R-B, Li X-T, Huang X. Topological organization of the brain network in patients with primary angle-closure glaucoma through graph theory analysis. Brain Topogr. 2024;37(6):1171–85.
7. Wu C-L. Creative tendency with brain network efficiency: a graph theory analysis. Think Skills Creat. 2024;53:101556.
8. Ma L, et al. Event causal relation extraction in brain Connectomics: a model utilizing weighted joint constrained learning. In: IEEE transactions on emerging topics in computational intelligence; 2024.
9. Ke M, et al. Brain functional network changes in patients with juvenile myoclonic epilepsy: a study based on graph theory and granger causality analysis. Front Neurosci. 2024;18:1363255.
10. Rock C, Alvino A, Christinal H, Chandy DA. Enhancing human brain activity through a systematic study conducted using graph theory and probability concepts on a hydar prehistoric organism. In: Computational intelligence and deep learning methods for neuro-rehabilitation applications. Academic Press; 2024. p. 257–77.
11. Nguyen B, et al. Accessing the topological properties of human brain functional sub-circuits in Echo State Networks. arXiv preprint arXiv. 2024:2412.14999.

Chapter 4
Advancing Brain Connectomics with Graph Neural Networks: Applications, Challenges, and Future Directions

Shake Ibna Abir, Shaharina Shoha, Nazrul Islam Khan, and Sarder Abdulla Al Shiam

4.1 Introduction

4.1.1 Background on Brain Connectomics

Connectomics, as it pertains to the brain, is a term for the mapping and comprehension of the network of neural connections, i.e., the brain [1]. Brain connectomics is critical in understanding how the brain functions, how cognition and behavior occur, via the underlying mechanisms. It permits researchers to examine how disparate brain locations collaborate and describe their parts in particular cognitive functions, like processing of memory, language, and sensory cognition.

In the sense of brain connectomics, the first form of brain connectivity is structural connectivity, which is discussed in Fig. 4.1 in all its three forms, which is the railway of the brain. Visualizing these networks then allows the researcher to begin to examine what the intricate relationship among the different regions of the brain really should like, and what each part of the "cognitive network" does in concert to give rise to cognition and behavior.

S. I. Abir (✉) · S. Shoha
Department of Mathematics & Statistics, Arkansas State University, Jonesboro, AR, USA
e-mail: sabir@astate.edu; sshoha@astate.edu

N. I. Khan
Department of Mathematics & Statistics, Stephen F. Austin State University, Austin, TX, USA
e-mail: khanni@jacks.sfasu.edu

S. A. Al Shiam
Department of Management, St. Francis College, New York, NY, USA
e-mail: sshiam@sfc.edu

M. M. Hassan et al. (eds.), *Graph Neural Networks for Neurological Disorders*,
https://doi.org/10.1007/978-3-032-04315-3_4

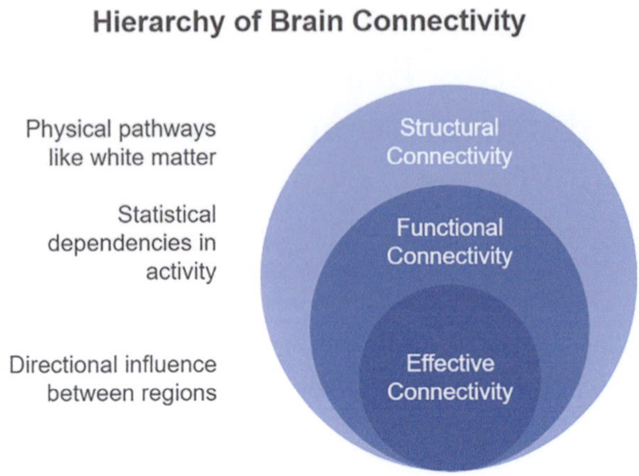

Fig. 4.1 Overview of brain connectomics

This figure gives a picture of brain connectomics, depicting different brain regions and their interrelation. The diagram illustrates the three types of brain connectivity. Fundamental to both understanding brain function and the organization of cognitive processes like memory, decision-making, and sensory perception are these different forms of connectivity.

Brain connectomics has great importance to understand how brain structure relates to brain function by mapping the dynamic organization of the brain, complex relationships between brain structure and brain function. In more recent times, recent advances in neuroimaging techniques over the past decade have enabled researchers to see these connections at a more detailed level, specifically enabling them to discover how local topological signatures reflect cognitive functions, identification of individuals, and behavior [2]. Therefore, brain networks provide a good means of exploring brain organization and function.

The history of the brain network development has in fact passed from localist brain function theories to distributed brain network theories. In the past, brain function was assumed to be localized in certain, specific areas of the brain, but eventually realized that the cognitive processes used different areas of the brain to interact [3]. By communicating through a network of these regions, they become capable of more sophisticated processing which is important for cognitive and other behavioral tasks [4].

The idea of connectomics unites ideas and problems from the fields of neuroscience, graph theory, computational modeling, and behavioral and cognitive experimentation to come to terms with how brain connectivity maps into behavior and cognition. The analysis of the brain network in graphs allows researchers to uncover the organizational properties of the brain in terms of small-worldness, modularity, and connectivity [5], properties that are essential to understand how the brain develops and carries out complex tasks in response to changing environments. These models also allow one to look at disease states, since changes in connectivity are associated with many cases of neurological disorders.

4.1.2 Importance of Studying Brain Networks

It is studied because the understanding of brain networks pertains to the understanding of these cognitive processes, behavior, and disease. The brain regions work together whose activity is necessary to coordinate functions of perception, decision making, and motor control. For example, addiction can be linked to broken areas in the brain's reward network and other parts of the brain which helps create and foster addictive behaviors [6]. Researchers can use comparison with these networks under different conditions to learn potentially how the mechanics underlie mental and neurological disorders. Brain networks are increasingly employed in clinical settings to understand the neurodegenerative disease of Alzheimer's and Parkinson's disease. It has been found that diseases of this class are typically related to specific brain network disruptions [7]. Because brain connectomics provides clinicians with a network-level way to understand these diseases, they can lead to feeling better able to diagnose and even target therapeutic strategies. Researchers are able to study what abnormally wired connectivity patterns exist in people with these conditions, allowing us to educate ourselves on the role that atypical brain network patterns play in these disorders.

4.1.3 Graph Theory in Neuroscience

Therefore, graph theory, whose origins lie in mathematics, is the perfect paradigm for modeling and analyzing the complexity of networks, as Maggioni essentially uses to do the same for networks in the brain. In graph theory, a graph is an entity, or again a set of entities together with edges (relationships among entities). Using graph theoretic techniques, researchers are able to quantify network properties like network connectivity, network efficiency, network resilience, and so forth related to the brain's architecture [8]. The brain's network can be analyzed on both the macro and micro scales as with this. As an example, fMRI data can be used to analyze macro-level brain networks (whole brain communication), whereas micro-level networks: (connections between the individual neuron or small clusters of neurons) can be evaluated using electrophysiological methods. By using this multi-level approach, researchers can study how brain networks work in both healthy and diseased conditions, and in doing so, they can find out how the abnormalities in the network structure and function cause cognitive impairment [9]. Secondly, graph theory makes it possible to analyze brain networks on different scales and resolutions, which is indispensable to understanding of brain function complexities. Graph theory has also been used as a tool to analyze biological systems and was used in other complex systems like aging and immune response [10]. These hub regions are prone to disruption and disruption in these regions may be correlated with neurological disorders, making them important targets for evaluation of novel therapeutics.

4.1.4 Graph Neural Networks (GNNs) in Neuroscience

Over the years GNNs have revolutionized application of the method in many fields, including neuroscience. GNNs were first used for graph-based tasks on social network analysis and recommendation systems, but over the years their task was applied for neuroscience [11]. For example, in network neuroscience, GNNs have been used to predict cognitive performance or to identify neurodegenerative disorder through disrupting brain network [12]. GNNs are able to deal with high-dimensional, complex data, which is ubiquitous in neuroimaging studies and can yield new revelations into the brain's operation. Given the power of GNNs, researchers are able to move up from traditional machine learning models and analyze brain networks directly in their natural graph structure. As a result, topological features underscoring the brain's connectivity patterns are extracted that are more representative of the brain's connectivity patterns. Unlike CNNs or fully connected networks (FNNs), which apply well to regular structure, GNNs are enabled to handle the irregular structure of brain networks. Therefore, GNNs can greatly enhance our knowledge of how the brain's elaborate connections work.

4.1.5 Research Objective and Scope

1. The goal is to investigate the application of GNNs in brain connectomics: Analyze what GNNs can do to model and analyze the brain networks and how GNNs could leverage their power to deal with complicated graph-structured data that appear in neuroimaging studies.
2. Concisely describe the advantages of GNNs over traditional methods of brain network analysis: Explain how using GNNs can be advantageous in brain network analysis, particularly their capability to represent non-Euclidean relationships and complicated topological features of brain networks.
3. Identify the key challenges that hinder the adoption of GNNs on the brain connectomics: namely, data complexity, computational costs, and limitation on the model's interpretability.
4. Investigate the potential of GNNs: this will help to understand and diagnose neurological disorders, such as neurological diseases using the power of data.
5. To provide insights into future directions in brain network research using GNNs: Offer recommendations for future research areas where GNNs can be integrated into brain connectomics, emphasizing their role in advancing the understanding of brain function, structure, and disease.

4.2 Theoretical Foundations of Brain Connectomics and Graph Theory

4.2.1 Introduction to Brain Connectomics

The scientific field of brain connectomics is what attempts to map brain structure and function. This gives a thorough understanding of the association of many areas of the brain, and the way they collaborate to work together, thus, to comprehend the normal functioning of the brain and the state of sicknesses or neurological disorders [13]. Brain connectomics helps understand how the information flows throughout the brain of a particular brain and this helps in understanding various cognitive functions. Within the context of connectomics, there are several types of brain networks that are analyzed. In structural connectivity, physical connectivity between brain regions is understood mainly through the white matter pathways. Finally, the terminology effective connectivity describes directional influence of one brain region on another (here, also known as Granger causality or dynamic causal modeling) [14]. A multi-dimensional view of the brain network architecture is provided by these three forms of connectivity that illustrate how the brain's structural organization is required for its functional dynamics. But how the brain's neural connectivity is organized as a network is important because the functioning of these networks is disrupted in a number of cognitive and behavioral disorders such as autism. For example, if there is a breakdown in functional connectivity, then the cognitive functions like memory and attention can be deficit, neurodegenerative diseases [15]. Therefore, brain connectomics is essential both for basic neuroscience and clinical applications for providing a framework to understand the brain's complexity and how it is changed in a disease.

4.2.2 Graph Theory Overview

Key concepts in graph theory include node degree, centrality, clustering, path length, and community structure. The degree k_i of a node i is the number of connections it has to other nodes, providing an indicator of the node's connectivity in the network. Betweenness and closeness centrality are examples from centrality measures that intended to find the most influential nodes in a network. Central nodes have high centrality and are pivotal for keeping communication across the network since they are anchors that connect between two parts of the network. Nodes tend to cluster (cluster) together to form small groups (small groups) or communities (communities) of very connected nodes. Community structure finds groups of nodes that are more closely connected among themselves than with other groups, and path length is the average number of edges between nodes that arc chosen at random in the graph [16]. Neuroscience uses different types of graphs depending on the nature of connectivity in the study. For instance, undirected graphs are naturally employed to represent brain networks with the connections between regions not having any direction. When there is some specific attention to the direction of influence of influence, directed graphs are used (e.g., for the causality in the context of effective connectivity models). In weighted graphs, strength of node's connections is taken into account while in bipartite graphs different types of nodes are considered, for example, brain regions and functional networks [17].

4.2.3 Graph Theory Applications in Neuroscience

Graph theory is a set of tools to investigate how the brain regions are organized and how the regions communicate in the brain as on large as a complex network. An interesting application of graph theory in neuroscience is in connection to neurodegenerative diseases which studies brain networks. Decreases in network efficiency and structural and functional connection loss when coupling between brain regions are observed in brain connectivity [18]. Researchers are able to use graph theory to analyze these changes of the networks and to detect potential biomarkers for these diseases and to improve diagnostics. Mapping the brain's connectivity using graph theory is one of the most ambitious of all initiatives, and one is that of the Human Connectome Project. To gather information on both structural and functional brain connectivity, this project aims to give a whole-brain network map on a detailed level. The results of the project have crucially paved the way for the advancement of our understanding of the brain structure and consisted the brain locales that act as centers within the network and contribute to the efficient working of the network [19]. Graph-based models have also been used in mapping the brain networks in the course of cognitive tasks as diverse as perception, memory, motor control, and have shed across different environments. Graph theory is also applied to the neural network modeling in the study of neuroscience, a task by which graph-based models are utilized to simulate the way in which neurons and brain regions interact. These models proved very useful in understanding how the brain works at microscopic as well as macroscopic levels. Therefore, graph-based models of neural networks have been used to simulate the way that information is communicated between neurons to understand the dynamics of brain activity under distinct cognitive states [20].

Brain networks are constructed per the graph theory and are visually presented in Fig. 4.2. This allows researchers to apply a range of graph theoretical techniques to quantify the organization and efficiency of brain networks through mapping of brain regions to nodes, and neural connections to edges. The value of neuroimaging data in building the brain networks is represented here, as it allows researchers to probe how certain types of connectivity make a contribution to brain function. Investigating healthy and disrupted brain activity of the brain therefore requires the understanding of these networks.

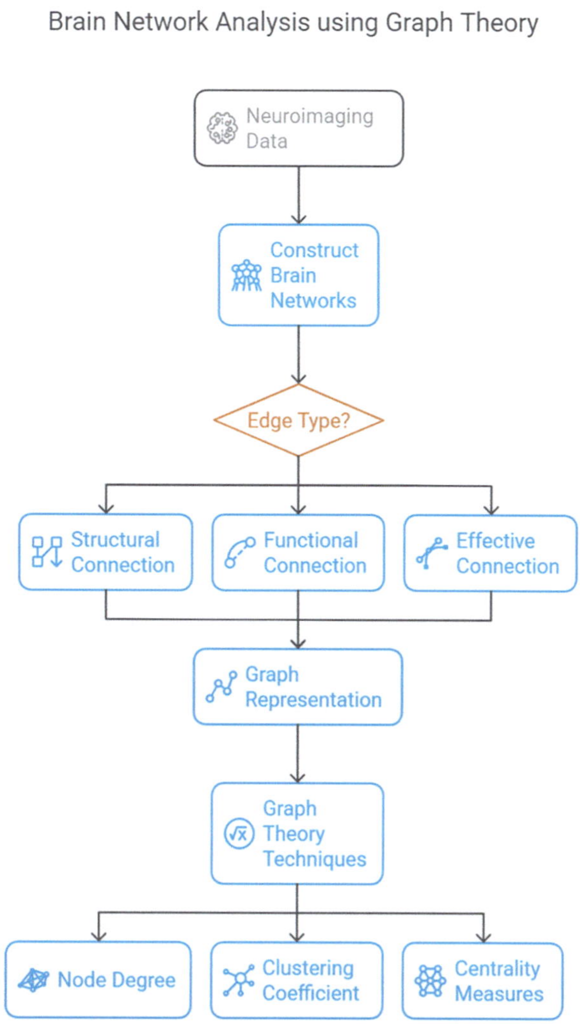

Fig. 4.2 Graph representation of brain networks

4.2.4 *Introduction to Graph Neural Networks (GNNs)*

Complex networks such as brain networks are the application for which these GNNs are very well suited. The biggest benefit GNN enjoys is the message passing between nodes and information is not limited to confined within an arbitrary node. GNNs learn rich node representations that encode the node features and their relationship with neighboring nodes by passing messages in this message passing process [21]. Figure 4.3 compares the network efficiency of brain regions in healthy individuals and patients with neurological disorders. Finally, network efficiency, which is the measure of how easy it is to transmit information in the brain between brain regions, is decreasing in these disorders, as the graph shows. These data support a role for disruption of brain connectivity in causing the cognitive and behavioral symptoms of these diseases.

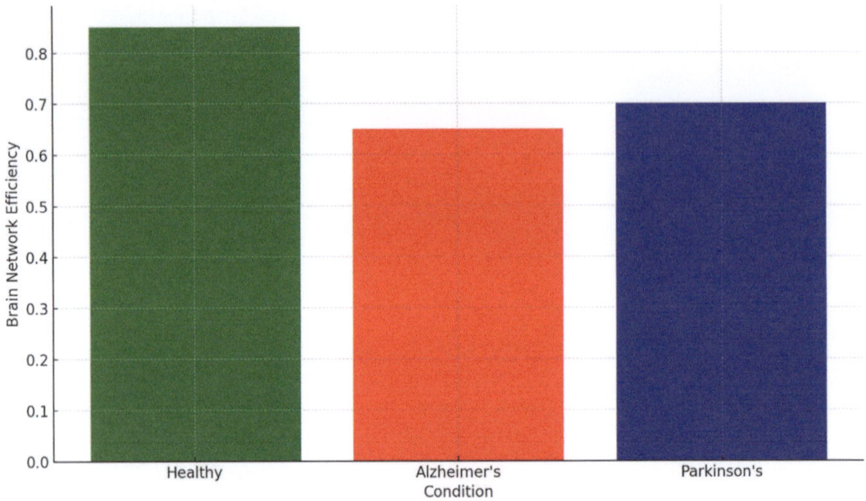

Fig. 4.3 Brain network efficiency in different neurological conditions

4.2.5 The Importance of GNNs in Brain Connectomics

Given their capability to deal with the complex programming of brain data, GNNs became an essential tool for analysis of brain networks. On the other hand, GNNs can directly work on graph data, learning the topological features which embody brain network [22]. GNNs are effective at brain connectomics as the ability to grasp the structural characteristics of brain networks is one of the GNN strongest suits. Applications of GNNs include predicting brain network dysfunction in Alzheimer and Parkinson's diseases in order to provide an earlier diagnosis and personalized treatment [23]. In addition, GNNs are extremely scalable and amenable to combinations with different types of neuroimaging data such as structural MRI, fMRI, or EEG. The ability to combine different data modalities makes it possible to analyze brain connectivity in much deeper perspectives, as this is related to behavior and cognition [24].

4.3 Brain Network Representation in Graph Theory

4.3.1 Types of Brain Networks

There are three fundamental sorts of brain networks based on the nature of connections between brain regions which are the functional connectivity structural connectivity and effective connectivity. There are different insights provided by each type, with respect to how networks in the brain operate interact.

Structural connectivity describes the physical connections between the regions of the brain in terms of their anatomical connections and hence is often estimated using neuroimaging techniques. The signal is the white matter tracts that connect the different areas of the brain. The brain's network is built upon structural connectivity that determines the pathway along which information may be conveyed between different regions [25]. For example, areas that are connected structurally, such as higher order regions of the prefrontal cortex that control higher order cognitive functions are connected to motor and sensory areas that participate in motor and sensory processing to ensure coordinated communication during complex tasks.

As for functional connectivity, the brain regions of interest are characterized by statistical dependencies in their activities. Functional connectivity represents the synchronization of activity and the degree to which one region appears to be in sync with another region. While functional connectivity does not necessarily mean a direct physical connection between the regions, it suggests temporal coordination between the regions [26]. For example, while being studied during a memory task, regions important to encoding, storing, and recalling information reveal high levels of functional connections with one another to do their work together.

By exploring the causal relationship of the ripple phenomena between brain regions, effective connectivity (further than effective connectivity) can be even pushed a step further. This is important to see the directional flow of information as well as how a region may modulate another region's activity to certain stimuli or tasks. Effective connectivity is often assessed by techniques such as dynamic causal modeling (DCM) or by Granger causality. Using these methods, researchers can build a model of brain network dynamics and can identify influences of activity in one region to others over time [27]. Knowing of effective connectivity provides a key to decipher the hierarchical structure of the brain in which some areas, like the prefrontal cortex that can influence or control the activity of others like sensory or motor areas during decision-making.

These three types of connectivity, together, give a complete picture of brain networks to bridge the gap between anatomical structures and cognitive functions. The cognitive functions have been integrated into a unified model of the brain far advanced the field of brain connectomics, laying the ground for understanding of complex brain function and their disruptions at the neurodevelopmental disorders.

4.3.2 Graph Construction in Brain Connectomics

Brain networks from neuroimaging data are built and represented as graphs by a number of steps: first, neuroimaging data is acquired; second, concatenated statistical maps of activations are created; third, delineation of brain locations or node locations is performed; finally, edges and node weights are defined using common statistics. The common process is to define brain regions (nodes), link regions (edges) and apply graph theory to the analysis of the properties of the resulting network.

Constructing methods of brain networks involve different neuroimaging modalities; however, fMRI is a very widely used technique in functional network construction, which measures the brain's hemodynamic response to neural activity and correlates time series of activity between various brain regions. EEG is frequently used to investigate dynamic functional connectivity of the brain based on phase and coherence of oscillatory brain activity across regions and has high temporal resolution. On the other hand, DTI has been used for structural connectivity by following the directions of white matter fibers to give information of anatomical connections connecting brain regions underlying the functional interactions between them [28].

After data collection by neuroimaging, the nodes and edges of brain network are defined. Nodes in brain connectomics are usually brain regions (certain cortical areas or sub-cortical structures). The connections themselves can be based on any structural, functional, or effective connectivity measures depending on the type of data used. The graph thus obtained can either be undirected or directed according to whether relationships between regions are symmetrical (as in structural connectivity) or directional (as in effective connectivity).

These networks are then analyzed using the graphtheoretic techniques. For instance, Graph Convolutional Networks (GCNs) [29] that are used to model the brain networks and extract features from a graph structure offer a useful way to process and classify network data.

4.3.3 Graph Representation Models

That is, a multitude of graph models are used in brain connectomics, all of which offer a different representation and analysis of brain network data. These models are based on a mathematical structure to define and manipulate the connectivity data, such as a matrix structure. The adjacency matrix is one of the most commonly used representations. Adjacency matrix is one of the fundamental elements of graph theory, where it can be used to quantify the connection strength of two nodes and provide findings via several network analyses, like calculating the node degree and relative network efficiency [30]. The degree matrix is the other important matrix. Understanding the overall connectivity of the network requires the degree matrix, which is also needed in computing a number of graph metrics such as centrality and clustering coefficients. In spectral graph theory, the Laplacian matrix is combined with the adjacency matrix in order to analyze the graph properties, namely connectivity and overall structure [31]. The Laplacian matrix is defined as:

$$L = D - A, \tag{4.1}$$

where L is the Laplacian matrix, D is the degree matrix, and A is the adjacency matrix. The eigenvalues of the Laplacian matrix characterize the differences between the adjacency and degree matrices and give an idea of the entire network efficiency and the ease of transmission of information between regions of the network.

Table 4.1 Types of brain network representations and their applications

Type of network	Data source	Application
Structural connectivity	DTI	Mapping anatomical pathways in the brain
Functional connectivity	fMRI, EEG	Studying brain activity patterns during tasks
Effective connectivity	fMRI, EEG	Analyzing causal relationships between regions

4.3.4 Challenges in Brain Network Modeling

Brain network modeling offers the perspectives of how the brain is organized and how it functions, but a number of challenges remain for its application. There is a big challenge in neuroimaging data of noise and variability. There exist various sources of noise in common neuroimaging techniques, like motion artifacts, scanner limitations, physiological noise, and so on. Finally, these sources can distort the data and obscure the true brain patterns of the network so that brain connectivity cannot be accurately modeled [32]. This issue could only be addressed by advanced preprocessing methods, e.g., motion correction, signal filtering, and data normalization, to guarantee the data fed into the network models is as reliable and accurate as possible. This is also challenging because brain networks are high high-dimensional and highly complex. There are millions of neurons and billions of synapses in the brain, resulting in networks that are high-dimensional to an extent such that they are uninterpretable by their direct analysis. However, such large-scale data may pose difficulty in efficiently handling with traditional graph theoretic approaches, and thus more computational methods need to be used, such as dimensionality reduction techniques and scalable algorithms. Finally, there is another layer of complexity given by brain network temporal dynamics. The brain's networks are dynamic and they change as a result of cognitive tasks, learning, and external stimulus. In this regard, the dynamic changes need to be modeled with techniques that can capture the variations in brain activity and connectivity with respect to time, for instance, dynamic functional connectivity methods or time-varying graph models [33]. Table 4.1 represents the types of brain network representations and their different applications.

4.4 The Application of Graph Neural Networks in Brain Connectomics

4.4.1 Principles of Graph Neural Networks (GNNs)

The main aim of GNNs is to derive node representations that incorporate node information collected from its neighboring nodes in the graph [34]. In such representations, not only do these features capture the features of one node at a time, but they also learn relationships between nodes [35]. GNNs perform update in an iterative manner on the graph, applying message passing mechanisms to update the node features in each layer.

4.4.2 Applications of GNNs in Neuroimaging and Connectomics

For brain network classification, prediction of cognitive performance, or disease diagnosis, Graph Neural Networks have been used in a great variety in neuroimaging and brain connectomics. GNNs have been mainly used in the context of brain network classification, for giving the task of classifying brain states based on functional connectivity patterns. To give an example, GNNs can be used to classify brain networks based on cognitive state, for example, between rest versus task performance through the analysis of patterns of functional connectivity on the brain [36]. The significance of this task lies in understanding how the brain adapts among different neural states.

Figure 4.4 gives a schematic representation of the GNN architecture to address brain connectomics. In particular, it shows that brain networks are processed via several rounds of processing layers in the GNN in which node features are updated according to information from its nearby nodes in a layer. The power of this architecture is especially apparent for brain network modeling because complex relationships among brain regions are very common in understanding cognitive functions and predicting neurological disorders. This approach allows researchers to harness GNNs for understanding the brain connectivity on a more profound level. A prime benefit of using GNNs for brain network analysis is their capability to combine multiple neuroimaging data types. For instance, researchers can combine structural and functional connectivity information into a single graph using Graph Convolutional Networks (GCNs) [37] and have a more complex representation of brain network.

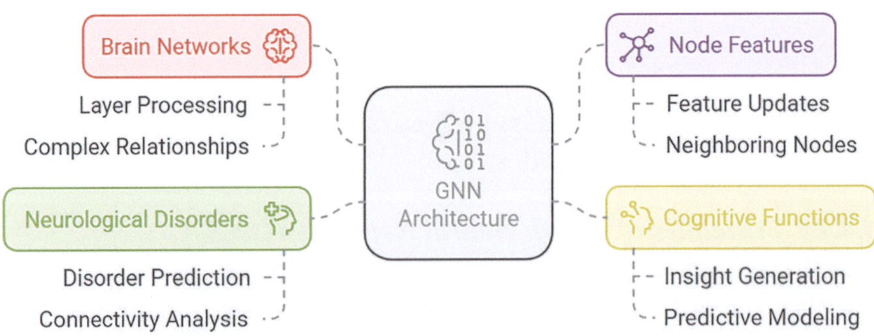

Fig. 4.4 GNN architecture for brain connectomics

4.4.3 *Graph-Based Deep Learning Approaches in Neuroscience*

There is a rise of popularity of deep learning approaches that combine GNNs with other models, e.g., CNNs and RNNs, for more complex tasks in neuroscience. A particular way of using the GNNs is with the recurrent neural networks (RNNs) to represent the dynamic brain networks. When remembering something or understanding language, temporal dynamics of brain networks are of interest in understanding how those networks were generated, and this approach is particularly useful to study those temporal dynamics in such tasks. Time spent in a given psychological state can therefore be captured not only through the temporal processing capabilities of an RNN, but also by combining the spatial representations from GNNs to reveal the structure of brain networks. A second example of integrating GNNs with deep learning is to employ CNNs to classify brain images while GNNs are used for analyzing the brain networks computed from these images. Using CNNs to extract high level feature from brain images and feed them to a GNN to analyze connectivity patterns between brain regions [38]. In tasks like classifying the brain regions associated with specific cognitive tasks or identifying the areas of interest to disease, this method has been used.

4.4.4 *Benefits of GNNs in Brain Network Analysis*

GNNs also prove to be a promising way of utilizing Graphs in brain connectomics, and several advantages in comparison to traditional machine learning methods are shown. The most important advantage is that they can retain the non-Euclidean relationships between the brain regions. Unlike CNNs, GNNs are created to deal with data in uneven graph-structured types [39]. Brain network analysis requires this because the brain's connectivity is inherently non-Euclidean; one region of brain connects to another, in a complex non-Euclidean fashion. Also, GNNs are able to scale efficiently to large datasets. Traditional machine learning techniques are computationally challenging to apply to brain network analysis since the data typically originate from large population and/or many subjects. However, GNNs can adequately handle such large–scale data as they process information in a distributed manner leading them to scale well with the rising size of the datasets [40]. It is particularly important with the scalability of this approach, which is particularly important in neuroimaging studies, in which datasets can consist of thousands of nodes and millions of edges. Apart from that, GNNs have a higher degree of interpretability than other black box models such as deep neural networks [41]. GNNs are able to identify some of the most influential key nodes and edges according to the model's predictions by focusing on the structure of the graph. It can provide a better understanding of the brain's organization and of the particular areas of the brain associated with different cognitive functions or diseases. In Fig. 4.5, GNNs are used

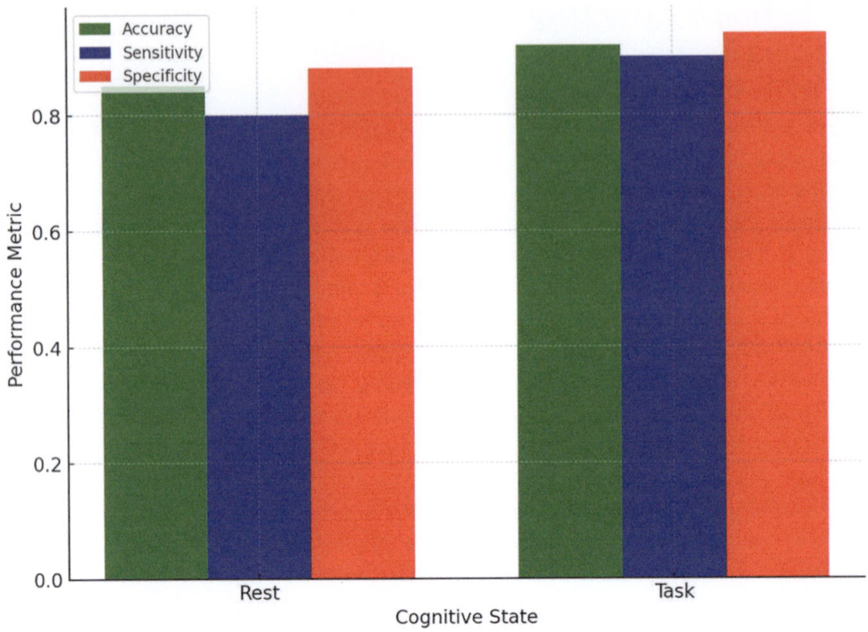

Fig. 4.5 Performance of GNNs in cognitive state classification

to classify brain networks by different cognitive states, e.g., rest versus task performance. The ability of GNNs to discriminate between cognitive states based on functional connectivity derived from fMRI scans is shown by evaluating the performance metrics (accuracy, sensitivity, specificity).

4.5 Methodology: Data, Preprocessing, and GNN Model Design

4.5.1 Data Acquisition and Sources

Accurate construction of brain networks requires that the quality of neuroimaging data be maintained. Some of these neuroimaging techniques are commonly used to acquire data on function and structure of the brain. fMRI, DTI, EEG, and sometimes PET are included [42]. For functional connectivity, fMRI is used mainly. It measures the changes in the neural activity usually referred to as Blood Oxygen Level Dependent [43]. The functional connectivity between these regions is defined in terms of the correlation between the time series between two of these places. Structural connectivity data is provided by DTI by following water diffusion orientation in white matter fibers and researchers can map the connections between brain regions [44]. These are the network's structural backbone. The brain's electrical activity can be measured at high temporal resolution in an EEG. In particular, it is useful for investigating dynamic functional connectivity or real-time brain activity in such a task.

4.5.2 *Preprocessing of Neuroimaging Data*

One important aspect of preprocessing steps is used to rid the neuroimaging data that is used to build brain networks of any artifacts that may occur. There are several steps to these and these steps vary across modality, but it generally consists of motion correction, normalization of space, artifact removal, and filtering.

4.5.2.1 fMRI Preprocessing

(a) Motion Correction: This step adjusts for any head movements during the scan.
(b) Spatial Normalization: Brain scans are transformed into a standard anatomical template (e.g., MNI space) to ensure accurate mapping of brain regions.
(c) Temporal Filtering: Low and high-frequency noise are removed to isolate the relevant neural signals.

4.5.2.2 DTI Preprocessing

(a) Artifact Removal: This step corrects for motion artifacts and eddy current distortions.
(b) Tractography: This step reconstructs the white matter pathways connecting brain regions. Once the data from these modalities is cleaned, it is then used to define nodes and edges. These relationships, whether functional, structural, or effective, form the backbone of the brain's network graph.

4.5.2.3 Constructing Brain Networks from Neuroimaging Data

Once the data has been preprocessed, the study constructs brain networks by representing the relationships between these regions as edges. Different types of brain connectivity are represented as graphs, depending on the type of neuroimaging data used:

(a) Functional Networks: These are constructed using time series data from fMRI or EEG, where edges between nodes are weighted.
(b) Structural Networks: These are built using DTI data, where the edges represent the strength or density of white matter tracts connecting different regions. The connectivity matrices constructed from these modalities are then thresholded to retain only the most significant connections.

4.5.2.4 Design and Implementation of GNN Models for Brain Networks

The design of GNN models for brain networks involves several stages, including the design of the model architecture.

Model Architecture

GNNs for brain network analysis typically involve multiple layers of graph convolutions. The basic update rule for a GNN at layer k can be represented as follows:

$$h_i^{(k)} = \sigma\left(W^{(k)} \cdot \sum_{j \in \mathcal{N}(i)} h_j^{(k-1)} + b^{(k)}\right), \tag{4.2}$$

where $h_i^{(k)}$ is the feature vector of node i at layer k, $\mathcal{N}(i)$ represents the set of neighboring nodes of node i, $W^{(k)}$ is the weight matrix for layer k, $b^{(k)}$ is the bias term, and σ is the activation function (commonly ReLU).

Graph Convolution

The graph convolution operation permits each hub to upgrade its highlight vector by amassing data from its neighbors. This handle is rehashed over numerous layers to capture progressively complex connections between hubs. The chart convolution operation can be generalized as:

$$H^{(k)} = \sigma\left(\hat{A}H^{(k-1)}W^{(k)}\right), \tag{4.3}$$

where $H^{(k)}$ is the node feature matrix at layer k, $\hat{A}$ is the normalized adjacency matrix of the graph, and $W^{(k)}$ is the weight matrix at layer k.

Loss Function

For classification tasks (e.g., predicting whether a subject has Alzheimer's disease), the cross-entropy loss is commonly used:

$$L_{CE} = -\sum_i y_i \log(p_i), \tag{4.4}$$

where y_i is the true label and p_i is the predicted probability for class i. For regression tasks, such as predicting cognitive scores, the mean squared error (MSE) is typically used:

$$L_{MSE} = \frac{1}{N}\sum_{i=1}^{N}\left(y_i - \widehat{y_i}\right)^2, \tag{4.5}$$

where N is the number of samples, y_i is the true value, and $\widehat{y_i}$ is the predicted value. Additionally, the model's generalization ability is assessed by evaluating its performance on unseen test data.

4.6 Results: Model Performance and Insights

4.6.1 Overview of Experimental Findings

GNNs have shown their power to improve the performance in brain connectomics, such as classification of the network, prediction in cognitive performance, and disease diagnosis. There are several studies that have shown that GNNs can make use of these features to predict outcomes ranging from cognitive states, presence of a neurological disease to even personalized treatment plans. Several experiments have been carried out to test the performance of the GNNs for the task of brain network prediction, such as in disease diagnosis and cognitive state classification. This work is important for understanding how dysregulation of the connectivity pattern in the brain is related to neurological and psychiatric diseases.

4.6.2 Performance of GNNs in Brain Network Prediction

High accuracy and robustness of GNNs in predicting some properties of brain network have been demonstrated. In these studies, a set of GNN models is built upon the so-called "brain network graphs" built using structural connectivity gotten from fMRI. They can learn complex relationship among brain regions, and find early biomarkers of AD that are often better than those found by traditional machine learning models. The accuracy of GNNs in disease diagnosis tasks is highly data set dependent and will depend on the complexity of the brain network features utilized. On this basis, GNNs get the accuracy of up to 90%, achieving large improvement compared to existing methods on studying AD. For example, GNNs have been recently applied for separating AD patients from health controls, with the achieved accuracy being 87% and high sensitivity and specificity. It results in both a key task in early diagnosis and intervention in neurodegenerative diseases and opens the doors towards new directions in diagnostic tools. GNNs have also been applied on other neurological conditions such as PD and schizophrenia. The GNN models presented their capability to well detect the disruptions of the basal ganglia–thalamocortical circuit crucial in PD pathology. This model has an accuracy of 85% better than traditional methods.

4.6.3 Comparison with Other Machine Learning Methods

It has been shown that GNNs possess unique advantages over other traditional machine learning models in brain network analysis, and their comparison has suggested that GNN is a promising tool for brain network analysis. Most of these models consider the brain data as a collection of independent features and do not incorporate the topological dependencies between brain regions. On the contrary, GNNs are tailored to deal with graph-structured data, which fits conveniently with analysis of brain networks whose relationship between nodes (brain regions) depends inherently on their connectivity. Therefore, GNNs have always been outperforming the traditional machine learning models, in the tasks like brain network classification and disease prediction. For instance, GNNs and SVMs were compared for prediction of AD, and the GNN model that was used got an accuracy of 87% whereas the SVM model only got an accuracy of 79%. Studies in Parkinson's disease and schizophrenia also showed similar improvements where GNNs were more accurate, more sensitive, and more specific than SVMs and random forests. In Fig. 4.6, it can be seen that GNNs and their variants predict if a person is infected with a neurological disease and also compare with other machine learning like SVMs and Random Forests.

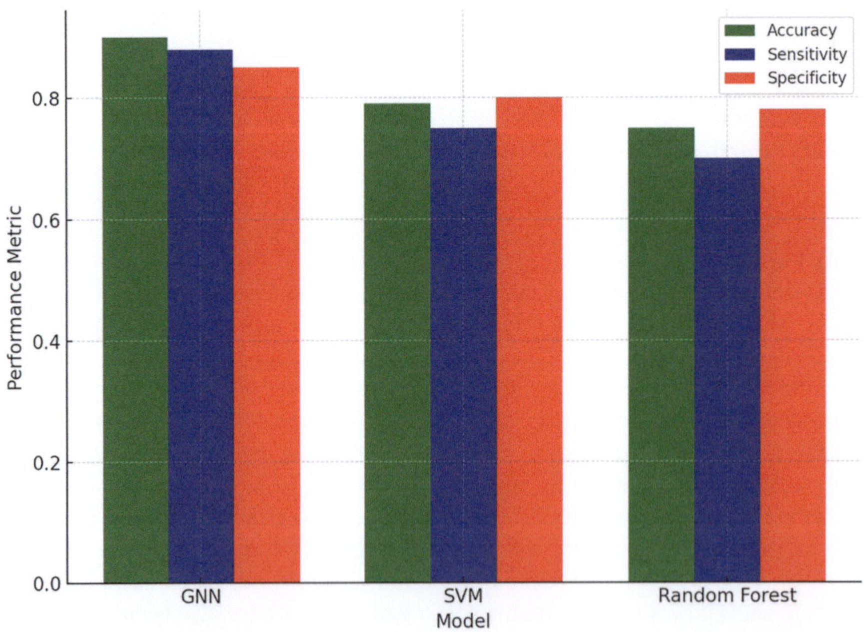

Fig. 4.6 Comparison of GNNs vs. traditional machine learning methods in disease prediction

4.6.4 Insights into Brain Network Properties

GNNs have been successfully applied to the analysis of brain network and enabled enlightening insights of brain's connectivity organization and dynamics. GNNs learn, from the graph structure of brain networks, to discover hidden patterns that are not so trivial to be seen by ordinary means. Insights derived from these models have direct relevance to understanding brain function and its dysfunction in a variety of cognitive and clinical contexts. Besides key regions finding, GNNs have also been employed to probe network topology in brain networks. It is observed that brain networks have small-world properties, where communities of neighboring regions exhibit high local cluster coefficients, and there exist short path lengths between different communities. This provides a quick way for the brain to communicate due to this efficient organization and it allows cognitive processes. Disruption of this small-world architecture has been associated with other neurological disorders such as AD and schizophrenia. In addition, GNNs have allowed to reveal temporal dynamics of brain networks. Temporal information from neuroimaging studies have been used to model brain networks with GNNs, the study uses these GNNs to model how brain networks change in time under cognitive tasks. For instance, GNNs were applied to analyze the memory recall study, where they could account for dynamic brain network reorganization when different parts of the brain are used to recall long-term memory. The dynamic view of brain networks allows one to see how different regions are coordinated to achieve such a task.

Fig. 4.7 represents that GNNs can capture temporal dynamics of brain networks. The diagram shows how the connectivity from brain regions (nodes) changes over time as a person engages in cognitive tasks. GNN layers aggregate information across time and each time step corresponds to a snapshot of the brain's network, such that changes of connectivity are represented. The figure highlights that GNNs leverage time-varying data to learn the dynamic interplay of different brain regions during such tasks as task-related memory recall, decision-making, or sensory processing.

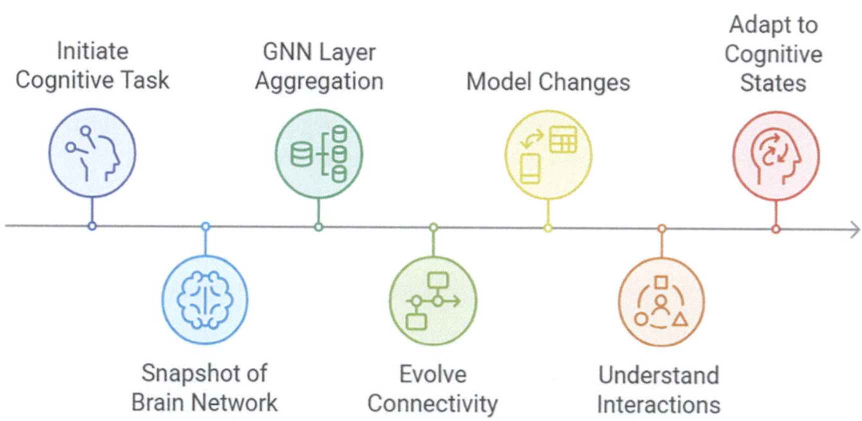

Fig. 4.7 Temporal dynamics in brain networks using GNNs

4.6.5 *Generalizability and Robustness of GNN Models*

One major advantage of GNNs in brain connectomics is that they can generalize across datasets and population. Unlike most traditional machine learning models that tend to overfit on certain dataset, GNNs can be trained to learn universal features in brain networks, which are generalizable to different data. In fact, this is especially important to carry out in neuroimaging studies, where data can be quite disparate in different populations, neuroimaging techniques, and scanning protocols. For instance, GNNs trained on one cohort of AD patients generalize quite well to another cohort from a different institution showing the model can learn universal features in the connectivity of the brain network. This generalizability is essential to apply GNNs to large-scale population studies and deploy such tools in clinical settings across populations of varied patients. While GNNs have much promise, it is important to note that there are also challenges associated with GNNs. The brain networks are extremely complicated and there is high variability in the data collected with neuroimaging techniques, making it a difficult problem to surpass a high accuracy threshold. Additionally, GNNs are trained with large amounts of labeled data which may limit studies with rare neurological disease or limited patient cohorts.

4.7 Discussion

4.7.1 *Implications of GNNs in Brain Network Research*

GNNs have lately arrived to the area of brain connectomics, allowing scholars to make a major leap forward in our ability to understand brain structure and perform. Traditional approaches for analyzing brain networks, like seed-based analysis, voxel-wise analysis, and simple graph theory, have given meaningful insights into brain function; however, while they are powerful in some ways they are limited a lot in capturing the complex and non-Euclidean relations that brain networks have [45]. Brain network analysis with traditional methods is usually done by treating brain regions as isolated units, while GNNs consider the dependency between the regions using the relationships present in the graph structure [46].

4.7.2 *Strengths and Limitations of GNNs in Brain Connectomics*

Advantages of GNNs are numerous, but so are its advantages and disadvantages when applied in brain connectomics, which need to be carefully considered.

(a) **Strengths:**

GNNs are able to directly deal with graph data, which is called the ability to capture non-Euclidean relationships [47]. Unlike traditional methods, which usually reduce brain connectivity to a gridded structure, GNNs can represent the brain's complicated, non-Euclidean topology. By this ability, GNNs can deduce more accurate and more rich representations of brain networks. These types of Machine Learning techniques are scalable in nature and can be used in large datasets with a large number of nodes and edges [48]. Especially for dealing with the huge amount of data that is common to neuroimaging-related studies, this scalability is important. GNNs enable integration of data from different types of neuroimaging modalities, including fMRI, DTI, EEG, etc. This multi-modal approach garnered us the chance to build more complete brain network models which take into account the structural and functional, and the effective connectivity. By doing this holistically, with consistent supervision of good direction, the accuracy and robustness of predictions is increased in disease classification tasks.

(b) **Limitations:**

I. Data Quality and Noise: As always, a challenging task in neuroimaging is to overcome inherent noise and variability in the data. A purpose of graph filtering methods is to reduce the effect of many artifacts, including motion, scanner variability, and physiological noise, which distort representations of brain networks in neuroimaging data [32]. Brain network data is high-dimensional resulting in thousands of brain regions and connections in the data. However, the problem with having so many features is that it can prove to be difficult to train models on them since it greatly increases the risk of overfitting [49]. To handle this, GNNs need large and diverse datasets to be able to generalize. Finally, since the feature spaces are quite high-dimensional, dimensionality reduction or regularization is frequently required to avoid overfitting.

II. GNNs: despite good predictive performance, are considered to be black boxes [50]. GNNs are of limited utility in clinical applications that demand interpretability due to this lack of interpretability, in particular for brain connectomics where it is essential to understand the biological significance of the brain network changes. To improve GNNs faith in clinical settings, efforts must be made to improve the explainability of GNN models, e.g., by visualizing important nodes or connections.

4.7.2.1 Potential Future Applications of GNNs

Due to the development of the field, GNNs have the potential to generalize to numerous applications in brain connectomics. There are some ideas of promising areas for future research and application. Current GNN models are mostly adopted for static brain network analysis; however, considerable effort has been paid to incorporate temporal dynamics of brain networks [51]. Over time, the brain's networks change, especially while you are thinking, feeling, or disease is progressing. Such a representation of brain function provides a dynamic, richer, more informative model of future GNN models that could integrate temporal information. Integration with Genetic and Environmental Data: An exciting future direction will be to integrate brain network data with genetic and environmental information. With GNNs, it would be possible to explore how genetic factors affect brain network connectivity and gain clues to genetic markers for different neurological and psychiatric disorders [52]. Besides, socioeconomic status or education as environmental factors could be incorporated into GNN models to look into how other outside factors can influence brain network organization and the resulting function. GNNs would be able to capture the variability in the brain networks of individual patients, which can be used to undertake personalized treatment plans for patients suffering from neurological disorders [53]. GNNs could thus predict how a given patient's brain network will respond to a given treatment, and could track a patient's brain network's response as the intervention progresses.

4.7.3 Integration with Other Techniques

Lastly, future study should investigate how GNN maps with other machine learning and deep learning techniques are integrated; e.g., how CNNs and reinforcement learning can be integrated, as well as if GNNs are related to unsupervised learning techniques. By combining these two approaches one may achieve still more robust models, applicable to the data from complex brain network. Also, multi-task learning could be incorporated in GNNs to predict multiple outcomes (e.g., disease progression, cognitive performance, and treatment response) given a network analysis of the patient's brain.

4.8 Conclusion

4.8.1 Summary of Findings

GNNs are a new type of graphs and networks and they have become a powerful tool for brain connectomics, providing a different way to analyze and understand brain networks. Particularly, these models are tailor-made for processing graph-structured data, and have been shown to perform very well in encoding the complex structure, such as that of connectivity patterns, between brain regions. Consequently, GNNs have seen great progress in investigating structural and functional brain networks in the past and a promising way to better understand the architecture and function of the brain as well as disease processes. These models have been effective in finding subtle changes in brain network connectivity that can lead to clinical symptoms, but could also be used in the early diagnosis and personalized treatment strategy. GNNs have further contributed to our knowledge about organization and dynamics of brain networks by elucidating important topological properties like network hubs, small world architecture, and modularity that lie at the heart of normal brain function and disease-induced disruption of these networks. Although GNNs have made impressive progress for brain connectomics applications, there are still many more challenges, including data quality, model interpretability, and scalability. Further, future research must address these challenges to fully realize the potential of GNNs to address problems in the clinic and in neuroscience.

4.8.2 Impact on Neuroscience and Clinical Applications

The utilization of GNN in brain connectomics has the possibility to change how well we comprehend the mind and its sicknesses. Conventionally, analysis of brain networks used simple models of brain connectivity, e.g., based on pairwise correlations or on a priori defined brain regions of interest. Although these methods have given very interesting information, they cannot capture some of the complexity of the brain network, which is a space with highly dimensional data and also with non-Euclidean relationships. While these intricate relationships may be too much for GNNs, they can be taught on the graph structure of brain networks which tends to result in more accurate models for how the brain functions or does not function. Researchers can train GNN models on large datasets made of neuroimaging data, and adopt these models to pinpoint which patterns of brain network disruption are characteristic of these diseases. It becomes possible to develop new early diagnosis and predictive modeling possibilities and treatment intervention before irreversible damage has set in. The second usage of GNNs is to track disease progression over time, since the GNNs can also tell how brain networks change as the disease evolves. In addition, GNNs are a tool for personalized medicine in neurodegenerative and psychiatric disorders. GNNs can predict how an individual will respond to certain treatments (pharmacologic or behavioral), by first analyzing a person's individual brain network. Using this approach, scholars could personalize treatment and, as a result, these treatments would be more effective for patients whose brain network profiles show they will respond to them; and less effective and cause fewer side effects in patients who will not benefit from them. GNNs also offer the advantage of processing multi-modal neuroimaging data. Multiple factors can play on brain networks and affect them, such as structural, functional, and metabolic changes that can cause cognitive deficits or disease pathology. Because GNNs are capable of merging multiple imaging modalities such as fMRI, DTI, and EEG, they can yield a more complete model of how the brain networks work. By taking a holistic approach, it gives a better understanding of how the brain is connected and how these connections link to behavior and disease, providing clinicians the ability to make choices on diagnosis and treatment with greater knowledge.

4.8.3 Future Directions in Brain Connectomics and GNNs

GNNs have already shown the potential in the brain connectomics however more development and investigation remain to be done. GNNs will have the increasing ability to address more complex questions about the brain and the disease using the neuroimaging techniques and the given datasets will get bigger and more diverse with time. An area of future work that seems to be promising is to combine temporal dynamics with GNN models. Brain networks change over time and especially in response to cognitive tasks, learning, and external stimuli. However, most of the current GNN models are static graph models that do not consider the dynamic property of brain networks. With additional temporal information, researches can incorporate the ideas of how the brain networks change in time, and thus create a more realistic depiction of the brain on different tasks and when it adapts to changes. In particular, this could be useful in studying other cognitive processes, like the brain connectivity changes leading to memory recall or decision-making processes. Regarding another area of growth, the development of explainable AI techniques for GNNs follows. Although GNNs can be very powerful tools, as predictors and classifiers, they tend to be by nature "black box" models, in the sense that they are not very interpretable. Understanding why a model is doing some prediction is critical to getting clinicians to trust a model in a clinical setting, especially when the model's familiar reasoning process is responsible for decisions as complex as diagnosing a disease, or playing a role in selecting a treatment. For future research, attention needs to be paid to the improvement of transparency and interpretability of GNN models such that clinicians can trust and interpret what is being generated by these models. This may entail creating methods to see the crucial nodes or connections of a brain network and thus determine which areas or paths have power in the model's determination. Another avenue of future research related to unsupervised learning in brain connectomics is also interesting. Without training data labels, GNN models applied without supervision could find new patterns in the brain network data. By this approach, new insights about the organization and function of the brain could be recognized that might not have seen otherwise because of lack of labeled data in the healthy population or in the rare neurological conditions. GNNs could be supplemented with these unsupervised learning techniques, like clustering or anomaly detection, to learn the unique patterns of connectivity that are linked to various cognitive functions or diseases. Last but not least, with deepening of GNN, it will be crucial to investigate how to combine GNN with other deep learning techniques, for example, CNNs and RNNs. When combined, GNNs could learn spatial as well as temporal aspects of brain networks and attain a more holistic understanding of brain function. This may be especially useful for studying evolving brain networks with respect to cognitive tasks or treatment.

4.8.4 Final Thoughts

Graph Neural Networks are proving themselves a powerful tool in studying brain connectomics giving a completely new way to model and study the complex networks of the brain. What is more, non-Euclidean data is effortless for them (i.e., easily handled) to take non-Euclidean data, integrate all kinds of neuroimaging modalities, capture the intricate connectivity patterns, all of this potentially opens new vistas to understand brain function, cognition, and disease. While GNNs will become an integral part of new findings in the field of brain connectomics and in clinical applications such as early disease detection, personalized treatment, and precision medicine. While there are still significant remaining challenges in this area such as those concerning data quality, model interpretability, and scalability, it is believed that these are some problems that could be solved by GNNs in the future and that GNNs have the potential to transform neuroscience and clinical practice. However, with ongoing research and development, GNNs offer such promise to provide further insights into the brain's connectivity and, as a result, better diagnostic tools, more effective treatments, and ultimately better results for the people who have the neurological and psychiatric disorders.

4.9 Data Availability

Upon reasonable request, the corresponding author will make the datasets and code used in this study public.

References

1. Santoro A, Battiston F, Lucas M, Petri G, Amico E. Higher-order connectomics of human brain function reveals local topological signatures of task decoding, individual identification, and behavior. Nat Commun. 2024;15(1):10244.
2. Abir SI, Shaharina Shoha M, Hossain M, Rahman SM, Saimon SI, Intiser Islam M, Mamun AI, Khan NI. Deep learning-based classification of skin lesions: enhancing melanoma detection through automated preprocessing and data augmentation. J Comput Sci Technol Stud. 2024;6(5):152–67. https://doi.org/10.32996/jcsts.2024.6.5.13.
3. Sultana N, Saimon SI, Islam I, Abir SI, Hossain MS, Al Shiam SA, Khan NI. Artificial intelligence in multi-disease medical diagnostics: An integrative approach. J Comput Sci Technol Stud. 2025;7(1):157–75. https://doi.org/10.32996/jcsts.2025.7.1.12.

4. Li Y, Wei Q, Adeli E, Pohl KM, Zhao Q. Joint graph convolution for analyzing brain structural and functional connectome. In: International conference on medical image computing and computer-assisted intervention. Cham: Springer; 2022. p. 231–40.
5. Shaharina Shoha, Abir, S. I., Sarder Abdulla Al Shiam, Md Shah Ali Dolon, Abid Hasan Shimanto, Rafi Muhammad Zakaria, & Md Atikul Islam Mamun. Enhanced Parkinson's disease detection using advanced vocal features and machine learning. J Comput Sci Technol Stud, 6(5), 113–128. doi:https://doi.org/10.32996/jcsts.2024.6.5.10, 2024.
6. Tolomeo S, Yu R. Brain network dysfunctions in addiction: a meta-analysis of resting-state functional connectivity. Transl Psychiatry. 2022;12(1):41.
7. Ciurea AV, Mohan AG, Covache-Busuioc RA, Costin HP, Glavan LA, Corlatescu AD, Saceleanu VM. Unraveling molecular and genetic insights into neurodegenerative diseases: advances in understanding Alzheimer's, Parkinson's, and Huntington's diseases and amyotrophic lateral sclerosis. Int J Mol Sci. 2023;24(13):10809.
8. Ortega A. Introduction to graph signal processing. Cambridge University Press; 2022. Seguin C, Sporns O, Zalesky A. Brain network communication: concepts, models and applications. Nature Rev Neurosci 2023;24(9):557–574.
9. Saimon SI, Islam I, Abir SI, Nigar Sultana M, Hossain S, Al Shiam SA. Advancing neurological disease prediction through machine learning techniques. J Comput Sci Technol Stud. 2025;7(1):139–56. https://doi.org/10.32996/jcsts.2025.7.1.11.
10. Cohen AA, Ferrucci L, Fülöp T, Gravel D, Hao N, Kriete A, Varadhan R. A complex systems approach to aging biology. Nature Aging. 2022;2(7):580–91.
11. Liu Z, Zhou J. Introduction to graph neural networks. Springer; 2022.
12. Sohail MN, Jiadong R, Irshad M, Uba MM, Abir SI. Data mining techniques for medical growth: a contribution of researcher reviews. Int J Comput Sci Netw Secur. 2018;18:5–10.
13. Bessadok A, Mahjoub MA, Rekik I. Graph neural networks in network neuroscience. IEEE Trans Pattern Anal Mach Intell. 2022;45(5):5833–48.
14. Faskowitz J, Betzel RF, Sporns O. Edges in brain networks: contributions to models of structure and function. Network Neurosci. 2022;6(1):1–28.
15. Abir, S. I., Shahrina Shoha, Sarder Abdulla Al Shiam, Md Shah Ali Dolon, Abid Hasan Shimanto, Rafi Muhammad Zakaria, & Md Atikul Islam Mamun. Deep neural networks in medical imaging: advances, challenges, and future directions for precision healthcare. J Comput Sci Technol Stud, 6(5), 94–112. doi:https://doi.org/10.32996/jcsts.2024.6.5.9, 2024.
16. Sarwer MH, Saha TR, Abir SI, Shaharina Shoha M, Hossain M, Sultana N, Saimon SI, Islam I, Hasan M, Al Shiam SA. EEG functional connectivity and deep learning for automated diagnosis of Alzheimer's disease and schizophrenia. J Comput Sci Technol Stud. 2025;7(1):82–99. https://doi.org/10.32996/jcsts.2025.7.1.7.
17. Abir SI, Shaharina Shoha M, Hossain M, Sultana N, Saha TR, Sarwer MH, Saimon SI, Islam I, Hasan M. Machine learning and deep learning techniques for EEG-based prediction of psychiatric disorders. J Comput Sci Technol Stud. 2025;7(1):46–63. https://doi.org/10.32996/jcsts.2025.7.1.4.
18. Miraglia F, Vecchio F, Pappalettera C, Nucci L, Cotelli M, Judica E, Rossini PM. Brain connectivity and graph theory analysis in Alzheimer's and Parkinson's disease: the contribution of electrophysiological techniques. Brain Sci. 2022;12(3):402.
19. Abir SI, Shoha S, Al shiam SA, Khan NI, Shimanto AH, Zakaria M, Arefeen SMS. Deep learning application of LSTM(P) to predict the risk factors of etiology cardiovascular disease. J Comput Sci Technol Stud. 2024;6(5):181–200. https://doi.org/10.32996/jcsts.2024.6.5.15.
20. Abir SI, Shoha S, Al Shiam SA, Saimon SI, Intiser Islam M, Mamun AI, Hossain MM, Rahman SM, Khan NI. Precision lesion analysis and classification in dermatological imaging through advanced convolutional architectures. J Comput Sci Technol Stud. 2024;6(5):168–80. https://doi.org/10.32996/jcsts.2024.6.5.14.
21. Khemani B, Patil S, Kotecha K, Tanwar S. A review of graph neural networks: concepts, architectures, techniques, challenges, datasets, applications, and future directions. J Big Data. 2024;11(1):18.

22. Cui H, Dai W, Zhu Y, Kan X, Gu AAC, Lukemire J, Yang C. Braingb: a benchmark for brain network analysis with graph neural networks. IEEE Trans Med Imaging. 2022;42(2):493–506.
23. Mohammadi H, Karwowski W. Graph neural networks in brain connectivity studies: methods, challenges, and future directions. Brain Sci. 2024;15(1):17.
24. Abir SI, Shoha S, Al Shiam SA, Uddin MM, Islam Mamun MA, Shamsul Arefeen SM. Health risks and disease transmission in undocumented immigrants in the U.S using predictive ML. In: 2024 Sixth international conference on intelligent computing in data sciences (ICDS). Morocco: Marrakech; 2024. p. 1–6. https://doi.org/10.1109/ICDS62089.2024.10756308.
25. Sohail MN, Ren J, Muhammad MU, Rizwan T, Iqbal W, Abir SI. Bio tech system, group covariates assessment on real-life diabetes patients by fractional polynomials: a study based on logistic regression modeling, English, Journal article, USA, 1944–3285. Edmond J Biotech Res. 2019;10:116–25.
26. Cao J, Zhao Y, Shan X, Wei HL, Guo Y, Chen L, Sarrigiannis PG. Brain functional and effective connectivity based on electroencephalography recordings: a review. Hum Brain Mapp. 2022;43(2):860–79.
27. Siddiqi SH, Kording KP, Parvizi J, Fox MD. Causal mapping of human brain function. Nat Rev Neurosci. 2022;23(6):361–75.
28. Prajapati R, Emerson IA. Construction and analysis of brain networks from different neuroimaging techniques. Int J Neurosci. 2022;132(8):745–66.
29. Li Y, Zhang X, Nie J, Zhang G, Fang R, Xu X, Li G. Brain connectivity based graph convolutional networks and its application to infant age prediction. IEEE Trans Med Imaging. 2022;41(10):2764–76.
30. Behrouzi T, Hatzinakos D. Graph variational auto-encoder for deriving EEG-based graph embedding. Pattern Recogn. 2022;121:108202.
31. Ko T, Choi Y, Kim CK. A spectral graph convolution for signed directed graphs via magnetic Laplacian. Neural Netw. 2023;164:562–74.
32. Avberšek LK, Repovš G. Deep learning in neuroimaging data analysis: applications, challenges, and solutions. Front Neuroimag. 2022;1:981642.
33. Mehedi MAA, Khosravi M, Yazdan MMS, Shabanian H. Exploring temporal dynamics of river discharge using univariate long short-term memory (LSTM) recurrent neural network at east branch of Delaware River. Hydrology. 2022;9(11):202.
34. Khoshraftar S, An A. A survey on graph representation learning methods. ACM Trans Intell Syst Technol. 2024;15(1):1–55.
35. Bhuyan BP, Ramdane-Cherif A, Singh TP, Tomar R. Convolution in neural networks. In: Neuro-symbolic artificial intelligence: bridging logic and learning. Singapore: Springer; 2024. p. 125–39.
36. S. I. Abir, S. Shoha, S. A. Al Shiam, M. M. Uddin, M. A. Islam Mamun and S. M. Shamsul Arefeen, "A comprehensive examination of MR image-based brain tumor detection via deep learning networks," 2024 Sixth international conference on intelligent computing in data sciences (ICDS), Marrakech, Morocco, pp. 1–8, doi:https://doi.org/10.1109/ICDS62089.2024.10756457, 2024.
37. Zhang L, Wang L, Zhu D, Alzheimer's Disease Neuroimaging Initiative. Predicting brain structural network using functional connectivity. Med Image Anal. 2022;79:102463.
38. Almohammadi, A. Graph convolutional neural network approaches for exploring and discovering brain dynamics. (Doctoral dissertation, University of Technology Sydney (Australia), 2024.
39. Jia Y, Wang J, Shou W, Hosseini MR, Bai Y. Graph neural networks for construction applications. Autom Constr. 2023;154:104984.
40. Zheng C, Chen H, Cheng Y, Song Z, Wu Y, Li C, Zhang S. ByteGNN: efficient graph neural network training at large scale. Proc VLDB Endow. 2022;15(6):1228–42.
41. Wang T, Zheng X, Zhang L, Cui Z, Xu C. A graph-based interpretability method for deep neural networks. Neurocomputing. 2023;555:126651.
42. Yen C, Lin CL, Chiang MC. Exploring the frontiers of neuroimaging: a review of recent advances in understanding brain functioning and disorders. Life. 2023;13(7):1472.

43. Lee QN, Chen JE, Wheeler GJ, Fan AP. Characterizing systemic physiological effects on the blood oxygen level dependent signal of resting-state fMRI in time-frequency space using wavelets. Hum Brain Mapp. 2023;44(18):6537–51.
44. Shamir I, Assaf Y. Tutorial: a guide to diffusion MRI and structural connectomics. Nat Protoc. 2024;1-19:317.
45. Miloserdov, K. Modelling the developing connectome: graph representation learning with variational autoencoders. (Doctoral dissertation, Wien), 2022.
46. Yan Y, Liu G, Cai H, Wu EQ, Cai J, Cheok AD, Fan Z. A review of graph theory-based diagnosis of neurological disorders based on EEG and MRI. Neurocomputing. 2024;128098:128098.
47. Hua C. Learning from graph-structured data: addressing design issues and exploring practical applications in graph representation learning. arXiv preprint arXiv. 2024:2411.07269.
48. Ma H, Rong Y, Huang J. Graph neural networks: scalability. In: Graph neural networks: foundations, frontiers, and applications; 2022. p. 99–119.
49. Aliferis C, Simon G. Overfitting, underfitting and general model overconfidence and underperformance pitfalls and best practices in machine learning and AI. In: Artificial intelligence and machine learning in health care and medical sciences: best practices and pitfalls; 2024. p. 477–524.
50. Bodria F, Giannotti F, Guidotti R, Naretto F, Pedreschi D, Rinzivillo S. Benchmarking and survey of explanation methods for black box models. Data Min Knowl Disc. 2023;37(5):1719–78.
51. Wein S, Schüller A, Tomé AM, Malloni WM, Greenlee MW, Lang EW. Forecasting brain activity based on models of spatiotemporal brain dynamics: a comparison of graph neural network architectures. Network Neurosci. 2022;6(3):665–701.
52. Xu Y, Zhang X, Li H, Zheng H, Zhang J, Olsen MS, Qian Q. Smart breeding driven by big data, artificial intelligence, and integrated genomic-enviromic prediction. Mol Plant. 2022;15(11):1664–95.
53. Paul SG, Saha A, Hasan MZ, Noori SRH, Moustafa A. A systematic review of graph neural network in healthcare-based applications: recent advances, trends, and future directions. IEEE Access. 2024;12:15145–70.

Chapter 5
GNNs for Neurological Disease Classification

Mahade Hasan, Farhana Yasmin, Yu Xue, and Md. Mehedi Hassan

5.1 Introduction

Millions of people worldwide are affected with neurological disorders like Alzheimer's, Parkinson's, and brain tumors, which constitute a serious and expanding global health issue. These diseases are often characterized by progressive deterioration of cognitive, motor, sensory functions, which severely impacts the quality of life and places a substantial burden on healthcare systems [1, 2]. The early and accurate classification of neurological diseases is critical for timely interventions, enabling more effective treatment and potentially slowing disease progression. Traditionally, the diagnosis and classification of these diseases have relied on clinical evaluations, including medical history and cognitive tests, as well as neuroimaging techniques such as magnetic resonance imaging (MRI) and computed tomography (CT) scans [3, 4]. These methods have limitations even if they offer helpful details about the structure and function of the brain. For instance, the interpretation of neuroimaging data is often subjective, relying heavily on the expertise of clinicians, which can lead to inconsistencies and potential misdiagnosis. Additionally, subtle early-stage disease markers are often missed, and changes in brain structure may not be readily apparent until the disease has progressed significantly. As neuroimaging technology advances, the volume and complexity of data have increased, creating a need for more efficient and objective methods to analyze and interpret these datasets [5]. DL and ML methods can process large volumes of data, uncover hidden patterns, and provide insights that might not be

M. Hasan (✉) · F. Yasmin · Y. Xue
School of Software, Nanjing University of Information Science and Technology, Nanjing, China
e-mail: mhasan@nuist.edu.cn; farhana@nuist.edu.cn; xueyu@nuist.edu.cn

M. M. Hassan
Computer Science and Engineering, Khulna University, Khulna, Bangladesh

M. M. Hassan et al. (eds.), *Graph Neural Networks for Neurological Disorders*,
https://doi.org/10.1007/978-3-032-04315-3_5

immediately apparent to human experts, improving diagnostic accuracy and enabling more personalized treatment strategies. Among these techniques, GNNs have shown considerable promise in healthcare applications, particularly in the classification and diagnosis of neurological diseases. GNNs are well-suited to handle data that is naturally represented as graphs, such as the complex and interconnected regions of the brain. In this representation, brain regions are treated as nodes, and the connections between them, whether structural, functional, or both, are modeled as edges [6]. This graph-based approach allows GNNs to capture the intricate relationships and dependencies between different brain regions, which is essential for understanding the dynamics underlying neurological diseases. Unlike traditional CNNs [7], which operate on grid-like data, GNNs can work with data that exists in non-Euclidean spaces, such as graphs, making them ideal for neuroimaging analysis. By processing brain network data in this way, GNNs can identify subtle patterns and changes in brain connectivity that may be indicative of disease. Furthermore, GNNs can integrate multimodal data, such as genetic information or functional imaging data, providing a more holistic view of the disease and improving classification accuracy. This ability to combine multiple sources of information is particularly valuable for diseases like Alzheimer's and Parkinson's, where environmental and genetic factors both contribute to the development of disease [8].

The ability of GNNs to model complex relationships and capture subtle changes in brain networks has significant implications for early disease detection, progression prediction, and biomarker discovery. By analyzing longitudinal data, GNNs can also help track how brain networks change over time, enabling the prediction of disease trajectories and the development of personalized treatment plans. Despite their potential, the application of GNNs to neurological disease classification is not without challenges. Issues such as data heterogeneity, limited availability of large-scale annotated datasets, and the high computational cost of training GNNs on neuroimaging data need to be addressed. Additionally, while GNNs are powerful at capturing complex relationships, the interpretability of these models remains an ongoing challenge. Understanding how GNNs arrive at their predictions is critical for their acceptance in clinical practice, where transparency and explainability are paramount. The objective of this chapter is:

- Explore the application of GNNs in the classification and diagnosis of neurological diseases.
- Discuss the potential of GNNs to revolutionize early disease detection, including identifying subtle markers that are often missed by conventional methods.
- Investigate how GNNs can contribute to personalized treatment strategies by incorporating multimodal data and understanding individual disease dynamics.
- Highlight the challenges in applying GNNs to neurological disease classification, such as data heterogeneity and computational complexity.
- Address the future directions for the integration of GNNs into clinical workflows, including the need for large-scale annotated datasets and improved model interpretability.

Through this exploration, the chapter aims to shed light on how GNNs can enhance diagnostic accuracy, improve patient outcomes, and contribute to the future of personalized neurology.

5.2 Fundamentals of Neurological Diseases

5.2.1 *Types of Neurological Diseases*

Neurological disorders include a diverse range of conditions that impact the brain, spinal cord, and nervous system, each characterized by unique causes and symptoms [9]. These conditions can be categorized into several types, as shown in Fig. 5.1.

Neurodegenerative diseases involve the progressive degeneration of nerve cells in the brain, leading to cognitive, motor, or behavioral impairments. Prominent examples include Alzheimer's disease, marked by memory loss and cognitive decline, Parkinson's disease, which impairs motor functions, and Huntington's disease, characterized by involuntary movements and cognitive dysfunction [10]. The gradual onset of symptoms in these diseases makes early detection difficult. Cerebrovascular diseases, such as stroke, occur when blood flow to the brain is disrupted, leading to sudden deficits in motor, speech, or cognitive functions. Strokes can be ischemic, caused by blocked blood vessels, or hemorrhagic, caused by bleeding in the brain. Brain tumors, whether benign or malignant, can interfere

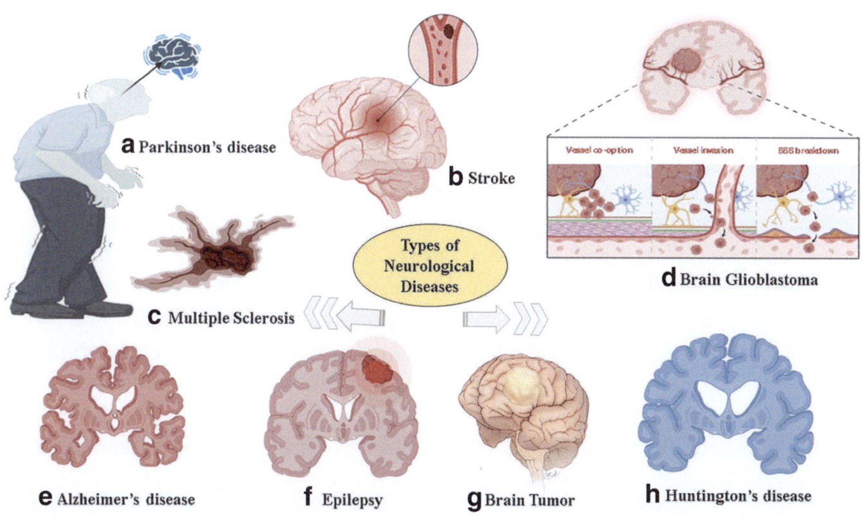

Fig. 5.1 Different types of neurological diseases: (**a**) Parkinson's disease, (**b**) multiple sclerosis, (**c**) stroke, (**d**) brain glioblastoma, (**e**) Alzheimer's disease, (**f**) epilepsy, (**g**) brain tumor, and (**h**) Huntington's disease

with brain function depending on their location. Common types include glioblastomas, meningiomas, and metastatic brain tumors, with symptoms such as headaches, seizures, and neurological impairments. Multiple sclerosis (MS) is an autoimmune disorder in which the immune system targets the protective sheath of nerve fibers, resulting in symptoms such as muscle weakness, vision impairment, and coordination challenges. Lastly, epilepsy and seizure disorders involve recurrent seizures due to abnormal electrical activity in the brain, which can result in convulsions, loss of consciousness, or unusual sensations. Each category presents unique challenges in diagnosis, progression, and treatment, emphasizing the need for advanced diagnostic methods and personalized therapeutic approaches.

5.2.2 Challenges in Diagnosis and Classification

The diagnosis and classification of neurological diseases face numerous challenges, largely due to the inherent complexity of the nervous system and the diverse nature of these disorders. One key challenge is the subtlety of early symptoms, as many neurological diseases, such as Alzheimer's and Parkinson's, have slow and gradual onset, often presenting with mild symptoms that may be mistaken for normal aging or other health conditions. These early-stage markers are difficult to detect using conventional diagnostic methods. Furthermore, the heterogeneity of disease progression adds another layer of complexity. Neurological diseases do not follow a uniform course across individuals; genetic, environmental, and co-existing health factors can significantly influence disease progression, making it difficult to establish generalized diagnostic criteria and complicating the interpretation of neuroimaging data.

Another challenge lies in the complexity of neuroimaging data. Techniques such as MRI and fMRI provide high-dimensional data that reflect both the structural and functional aspects of the brain. However, analyzing these complex datasets requires advanced computational methods, as traditional approaches often fail to capture subtle patterns or process the large volumes of data efficiently. Additionally, the lack of large annotated datasets hampers the development and training of ML models, as obtaining sufficiently large and labeled datasets is often difficult due to privacy concerns, small sample sizes, and the intricacies of accurate diagnosis.

There is also variability in neuroimaging modalities, such as structural MRI, fMRI, and PET scans, which each provide complementary yet distinct information. Integrating and interpreting data from these varied sources into a unified diagnosis is a complex task. Lastly, the subjectivity in diagnosis poses significant challenges. Traditional methods heavily depend on clinicians' expertise, which may introduce inconsistencies, especially in rare or complex neurological conditions.

To address these challenges, advanced computational techniques like GNNs offer significant potential. GNNs can capture complex relationships in brain networks by representing the brain as a graph, where nodes represent regions and edges represent connections. This approach allows the model to process complex, heterogeneous neuroimaging data more effectively. For instance, the mathematical representation of a GNN can be described as:

$$H^{(k+1)} = \sigma\left\{D^{(-1/2)}.AD^{(-1/2)}.H^{(K)}.W^{(k)}\right\}, \tag{5.1}$$

where $H^{(k)}$ represents the feature matrix at layer k; A is the adjacency matrix, D is the degree matrix, σ is an activation function, and $W^{(k)}$ is the weight matrix at layer k. Equation (5.1) allows GNNs to capture complex patterns in brain connectivity and improve the accuracy of early diagnosis, offering a promising solution to the challenges faced by traditional methods.

5.3 Graph Representation in Neurological Data

5.3.1 Using Graphs to Represent Neuroimaging Data

Graph-based approaches have gained considerable attention in the analysis of neuroimaging data due to their ability to capture the complex relationships between brain regions. Neuroimaging techniques such as structural MRI, functional MRI (fMRI), and diffusion tensor imaging (DTI) provide detailed information about the brain's structure and function. In the graph representation of the brain, the brain is modeled as a graph $G = (V, E)$ where:

***Nodes** (V):* Each node represents a distinct brain region or voxel. These regions are often defined based on anatomical regions of interest (ROIs) or functional networks.

***Edges** (E):* Edges between nodes represent connections between brain regions, such as structural, functional, or effective connectivity. The strength of an edge e_{ij} can correspond to the degree of connectivity between two regions i and j, and is often quantified by a weight w_{ij}, representing the intensity of the connection between nodes.

Figure 5.2. provides examples of modeling a brain network: (a) Structure of brain network, showing the anatomical connections between different brain regions, and (b) a weighted graph, where nodes represent brain regions and edges represent the strength of connections, quantifying the relationship between these regions in terms of connectivity.

This graph structure can be expressed mathematically as a weighted graph $G = (V, E, W)$, where W is the weight matrix that holds the values of the edge weights, w_{ij} represents the weight of the edge between nodes i and j.

By representing the brain as a graph, each individual's neuroimaging data can be treated as a unique graph, where node attributes w_i and edge weights w_j vary across subjects. This representation allows the use of graph-based techniques, such as **GNNs**, to model the complex interactions between different brain regions. GNNs

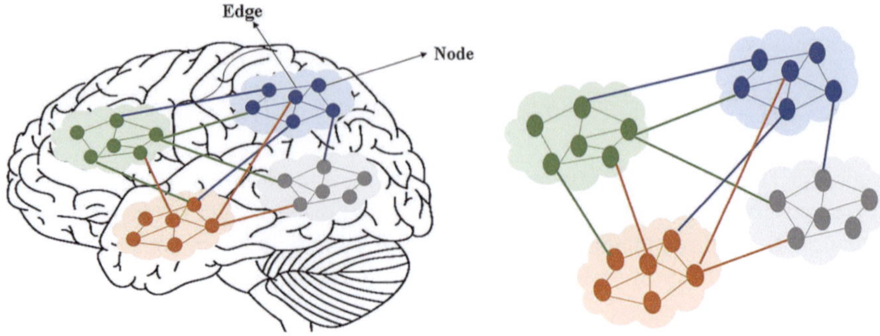

Fig. 5.2 Examples of modeling a brain network (**a**) structure of brain network, (**b**) weighted graph

are particularly suited for analyzing non-Euclidean data like graphs. The propagation of information through a GNN layer can be formulated as:

$$H^{(k+1)} = \sigma\left(\hat{A}H^{(k)}W^{(k)}\right), \tag{5.2}$$

where $H^{(k)}$ is the feature matrix at layer (k) representing the hidden features of the nodes, $\hat{A}$ is the normalized adjacency matrix, σ is the activation function, and $W^{(k)}$ is the weight matrix for layer (k).

This graph-based representation allows the model to capture both local and global patterns in brain connectivity, which is particularly useful for identifying biomarkers for neurological diseases. By detecting changes in connectivity patterns, GNNs can highlight alterations that may correlate with disease states, even before visible symptoms appear.

5.3.2 *Brain Network Construction for Disease Analysis*

Brain network construction is a critical step in analyzing neuroimaging data for disease diagnosis and understanding brain pathology. The process typically involves creating a network (graph) that reflects the functional and/or structural relationships between various regions of the brain. This network can then be analyzed for abnormalities associated with neurological diseases.

In Fig. 5.3, neuroimaging data are used to estimate functional or structural connectivity. Parcellations from structural MRI define brain regions as nodes. The adjacency matrix represents the relationships between nodes, forming the brain network, and graph theory metrics are applied to analyze the network's topology.

There are two primary approaches to constructing brain networks:

(a) ***Functional Brain Networks:*** In fMRI, the brain's activity is recorded over time, capturing temporal fluctuations in brain activity [11]. Functional connectivity describes the statistical relationships or correlations in activity among

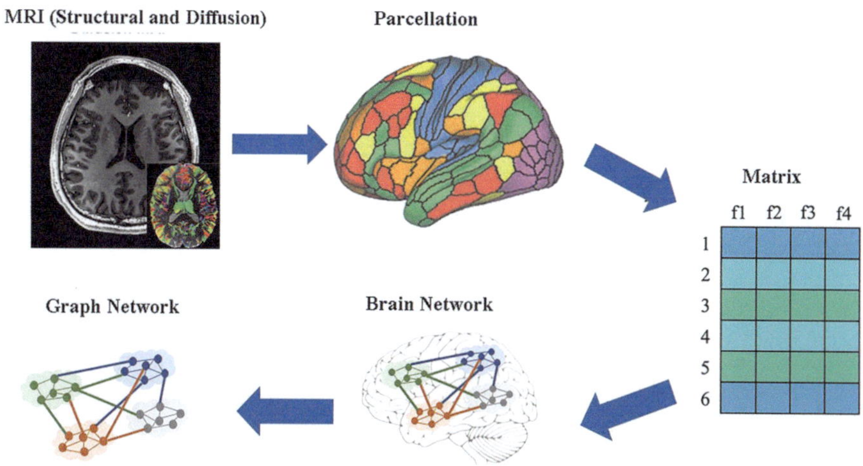

Fig. 5.3 Brain network construction schematic

various brain regions. Regions with highly correlated activity are considered functionally connected, forming an edge in the graph. Functional networks can be built by calculating correlation or coherence between the time-series data [12] of brain regions, and these networks are often dynamic, reflecting how connectivity patterns change over time. Disease-related changes in functional networks can manifest as alterations in the strength or topology of connectivity, which can be detected using advanced analytical techniques such as GNNs.

(b) ***Structural Brain Networks:*** Structural brain networks are based on the physical connections between brain regions, derived from techniques like DTI, which measures the direction and integrity of white matter pathways [13]. These pathways form the structural connections (edges) in the brain network. In a graph representation, the brain regions are represented as nodes $V = \{v_1, v_2, \ldots . v_n\}$, and the edges $E = \{e_{ij}\}$ represent the white matter tracts that connect these regions.

The strength of an edge e_{ij} is often quantified by a measure such as fractional anisotropy (FA), which reflects the integrity of white matter fibers between regions i and j. FA values range from 0 to 1. The weight of the edge w_{ij} is therefore defined as:

$$w_{ij} = FA(i,j), \tag{5.3}$$

where $FA\ (i,j)$ represents the fractional anisotropy value between the regions i and j.

In the structural brain network, alterations in the connectivity between brain regions, such as the loss of white matter integrity or disruption of white matter tracts, are commonly observed in neurological diseases like multiple sclerosis and Alzheimer's disease [14]. These disruptions can be modeled using graph-based methods to capture and quantify disease-specific patterns. For instance, the global connectivity of the brain, characterized by measures like global efficiency E_{global}, can be used to assess overall network integrity:

$$E_{\text{global}} = \frac{1}{n(n-1)} \sum_{i \neq j} \frac{1}{d_{ij}}, \tag{5.4}$$

where d_{ij}, the shortest path, is the distance between nodes i and j in the graph, and n is the total number of nodes in the network. Disruptions in structural connectivity, such as decreased FA values or altered network efficiency, can serve as biomarkers for neurological conditions, and these changes can be effectively detected and analyzed using graph-based methods like GNNs.

5.3.3 Hybrid Brain Networks

Hybrid brain networks integrate both structural and functional connectivity to provide a holistic understanding of brain function and integrity. In this framework, brain regions serve as nodes, while edges represent both structural and functional connections. The combined edge weight w_{ij}:

$$w_{ij} = a.w_{ij}^{\text{structural}} + \beta.w_{ij}^{\text{functional}}, \tag{5.5}$$

where α and β are weights that balance the contribution of each modality. By combining structural and functional data, hybrid networks provide a more robust representation of brain connectivity, which is especially useful in diseases like Alzheimer's, where both structural degradation and functional disruptions occur. This integration allows GNNs to identify subtle patterns indicative of neurological diseases, improving disease classification accuracy and offering deeper insights into disease mechanisms.

5.3.4 Disease-Specific Network Features

Brain network features, such as modularity and small-world properties, provide critical insights into neurological diseases. Healthy brains typically exhibit small-world properties, balancing high local clustering with short global paths for efficient information transfer. Neurological diseases like Alzheimer's disrupt these features, leading to fragmented or less efficient network topologies.

(a) ***Modularity*** *(Q):* Modularity quantifies how well a network can be partitioned into distinct modules or clusters. It is defined as:

$$Q = \frac{1}{2m}\sum_{i,j}\left[A_{ij} - \frac{k_i k_j}{2m}\right]\delta\left(c_{i,}c_j\right), \tag{5.6}$$

where A_{ij} Adjacency matrix (connection strength between nodes i and j), k_i *and* k_j Degrees of nodes i and j, m Total number of edges, $\delta(c_{i,}c_j)$ Indicator function, 1 if nodes i and j belong to the same module, and 0 otherwise.

(b) ***Small-World Properties:*** Small-world networks are characterized by a high clustering coefficient (C) and a short characteristic path length (L), enabling efficient information processing. These are computed as:

$$C = \frac{1}{n}\sum_{i=1}^{n}\frac{2e_i}{k_i\left(k_i - 1\right)}, \tag{5.7}$$

where e_i Number of edges between neighbors of node i, and k_i Degree of node i.

$$L = \frac{1}{n\left(n-1\right)}\sum_{i\neq j}d_{ij}, \tag{5.8}$$

where d_{ij} Shortest path between nodes i and j.

GNNs leverage these features to learn representations of brain networks. By modeling node embedding that incorporate modularity, clustering, and path length metrics, GNNs enhance the classification of neurological diseases. This method not only improves early detection but also aids in tracking disease progression and predicting treatment outcomes.

5.4 GNN Architectures for Neurological Disease Classification

5.4.1 *Overview of GNN Architectures*

GNNs are specialized neural networks that process graph-structured data by extending the concepts of CNNs and graph embedding techniques. While CNNs excel in image classification by working on grids of pixels, GNNs generalize these concepts to graphs, where the structure is defined by nodes and edges. GNNs are used for tasks like node classification, edge prediction, and graph-level classification [15, 16].

The input to a GNN is a graph $G = (V,E)$ where each node $v_i \in V$ has a feature vector v_i, and each edge $(v_i, v_j) \in E$ also have a feature vector e_{ij}. The process begins by encoding the graph structure into an embedding. The node feature vectors are iteratively updated through neural network layers by aggregating information from neighboring nodes. At each layer l, the feature vector $h_i^{(l+1)}$ of node v_i is updated as:

$$h_i^{(l+1)} = \sigma\left(\sum_{j\in N(i)} \alpha_{ij} W^{(i)} h_j^{(l)}\right), \tag{5.9}$$

where $N(i)$ is the set of neighbors of v_i; α_{ij} is an attention weight or normalization factor, $W^{(l)}$ is the weight matrix for layer l, and σ is an activation function.

For text-based tasks, GNNs treat each word as a node, forming a graph from sentence structures. RNNs handle sequential dependencies, but GNNs offer better performance for graphs with arbitrary size and complex structures, overcoming CNNs' limitations in these domains. This framework ensures information from nodes, edges, and the global graph context is preserved throughout the network [17].

In Fig. 5.4, the GNN architecture and processing procedure are illustrated. Initially, the GNN selects neighboring nodes based on a predefined strategy. Then, an aggregation function gathers information from these surrounding nodes. Finally, the aggregated data is processed through a neural network using a nonlinear transformation, producing an updated representation of the central node.

GNNs are tailored for processing graph-structured data, emphasizing both the relationships between entities and their individual properties. This makes GNNs particularly suitable for modeling the complex interactions between brain regions in neuroimaging data.

(a) ***Graph Convolutional Networks (GCNs):*** GCNs are a type of neural network that applies convolution-like operations directly to graph-structured data. They extend the concept of convolution from regular grids to arbitrary graph structures [18]. The fundamental principle involves iteratively updating node feature representations by aggregating information from neighboring nodes.

Let $G = (V, E)$ represent a graph, where V is the set of nodes and E is the set of edges. Each node $v_i \in V$ is associated with a feature vector $x_i \in R^F$, where F is the number of input features per node. The graph is described by its adjacency matrix $A \in R^{N\times N}$, where $N = |V|$ is the number of nodes. The degree matrix D is a diagonal matrix with $D_{ii} = \sum_j A_{ij}$.

The GCN layer updates the feature representation of nodes as:

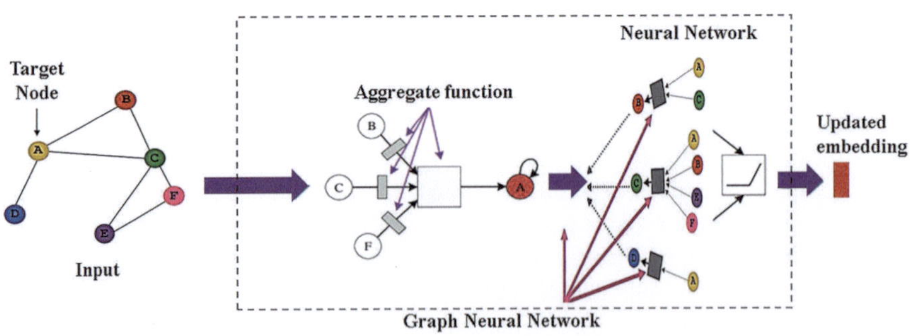

Fig. 5.4 A basic architecture of graph neural network (GNN)

$$H^{(l+1)} = \sigma\left(\hat{A}\right)H^{(l)}W^{(l)}, \tag{5.10}$$

where $H^{(l)} \in R^{N \times F_l}$: Node features at layer l, with F_l as the feature dimension; $W^{(l)} \in R^{F_l \times F_{l+1}}$; Trainable weight matrix for layer l; σ Activation function; $\hat{A} = \tilde{D}^{-1/2}\tilde{A}\tilde{D}^{-1/2}$: Symmetrically normalized adjacency matrix, where $\tilde{A} = A + 1$ and $\tilde{D}$ is the degree matrix of $\tilde{A}$.

For a multi-layer GCN with L layers, the node features are computed iteratively, starting with the input features $H^{(0)} = X$. The final output $Z = H^{(L)}$ represents the learned node embeddings, these updated node feature representations can be utilized for downstream tasks such as node classification, link prediction, or graph-level classification.

A graph with node labels Y, a two-layer GCN for node classification can be written as:

$$Z = \text{softmax}\left(\hat{A}\,\text{ReLU}\left(\hat{A}XW^{(O)}\right)W^{(1)}\right), \tag{5.11}$$

where X is input feature matrix, $W^{(0)}$ *and* $W^{(1)}$ trainable weight matrices for the first and second layer; Z is the output probabilities for each class.

This framework allows GCNs to effectively learn representations of nodes by leveraging graph structure and feature information.

Figure 5.5. explains the basic architecture of Graph Convolutional Networks (GCNs), showcasing how GCNs perform convolution-like operations on graph-structured data. The figure highlights the iterative update process of node features by aggregating information from neighboring nodes, enabling the model to learn from graph relationships.

(b) ***Graph Attention Networks (GATs):*** GATs introduce a dynamic attention mechanism to GNNs, where the importance of neighbors is learned through

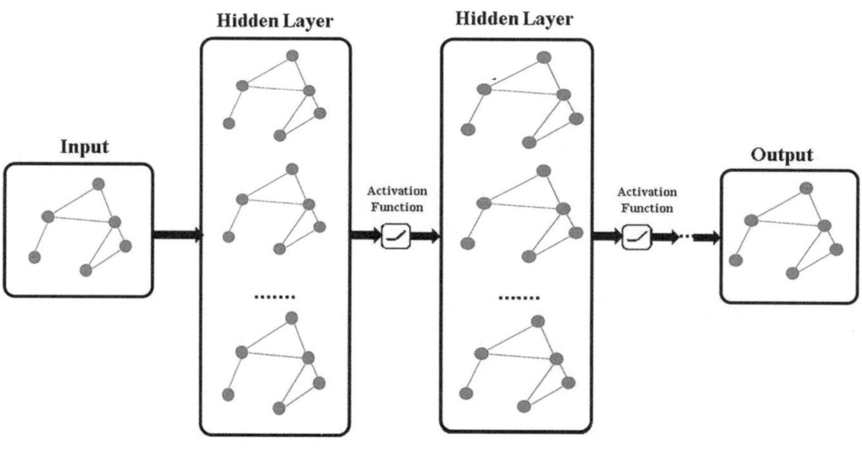

Fig. 5.5 Basic graph convolutional networks architecture

attention coefficients. For each node v_i, the attention coefficient a_{ij} between node v_i and its neighbor v_j is computed using a shared attention mechanism. The feature of node v_i is updated by aggregating its neighbors' features, weighted by a_{ij}. This approach allows GATs to focus on more relevant neighbors, improving performance on graph-based tasks. Multi-head attention is often used to stabilize learning by averaging or concatenating outputs from multiple attention heads.

(c) ***Graph Sample and Aggregation (GraphSAGE):*** GraphSAGE enhances the scalability of GCNs by sampling a fixed number of neighbors for each node, rather than using all neighbors. For each node, a subset of neighbors N_i is sampled, and their features are aggregated using a function like mean, LSTM, or pooling. The aggregated feature is then passed through a non-linear activation function to update the node's representation. This method reduces computational complexity and allows GraphSAGE to handle large graphs effectively while learning meaningful node embeddings.

(d) ***Graph Isomorphism Networks (GINs):*** GINs improve upon traditional GCNs by using a powerful aggregation function that captures graph structures more effectively. GIN uses a sum-based aggregation, where the features of node v_i and its neighbors are summed and passed through a MLP to update the node's representation. The injective aggregation ensures that GIN can distinguish non-isomorphic graphs, making it highly expressive and suitable for tasks like graph classification. The MLP introduces non-linearity, enabling the model to learn and capture more intricate relationships.

These architectures provide the foundation for more advanced models tailored to address the unique challenges of neurological disease classification.

5.4.2 Specialized GNN Models for Disease Prediction

While traditional GNN architectures provide a robust starting point, specialized GNN models have been developed to enhance their performance in the context of neurological disease prediction. These models often incorporate domain-specific knowledge and modifications to better handle the characteristics of neuroimaging data.

(a) ***Brain Network-Specific GNNs:***

In brain network-specific GNNs, the nodes V represent brain regions and the edges E represent functional or structural connections. The adjacency matrix A encodes these connections, with $A_{ij} = 1$ if there is a connection between regions i and j, and 0 otherwise. The node features X are associated with brain regions, often including attributes such as activity or structural metrics.

The GNN model aggregates neighborhood information for each node i by applying an update rule:

$$h_i^{(l+1)} = \sigma\left(W^{(k)}.\left(h_i^{(k)} + \sum_{j\in N(i)} A_{ij} h_i^{(k)}\right)\right), \tag{5.12}$$

where $h_i^{(k)}$ is the feature vector for node i at the k layer; $N(i)$ denotes the neighbors of node i; $W^{(k)}$ is the learnable weight matrix at layer k; σ is an activation function.

For disease-related tasks, such as predicting Alzheimer's, the model may incorporate prior knowledge of affected regions by giving higher weight to certain nodes or edges, guiding the GNN to focus more on those regions in the aggregation process. The output can then be used for classification or regression tasks, for example:

$$\hat{y} = \text{Softmax}\left(W^{(L)} h_i^{(L)}\right), \tag{5.13}$$

where L is the final layer and $\hat{y}$ is the predicted label.

(b) ***Hybrid Models:***

Hybrid GNN models integrate the advantages of CNNs and GNNs, allowing them to utilize both spatial and temporal features in neuroimaging data. Let X_{CNN} represent the feature map extracted by the CNN from an MRI scan, and $\boldsymbol{G} = (\boldsymbol{V}, \boldsymbol{E})$ represent the graph of brain regions and functional or structural connections. The CNN first extracts local features $f_{CNN} = CNN(X_{MRI})$. Then, the features are passed into the GNN for global relationship modeling:

$$h_i^{(l+1)} = \sigma\left(\sum_{j\in N(i)} \frac{1}{c_{ij}} W h_j^{(l)} + f_{CNN}\right), \tag{5.14}$$

where $N(i)$ represents the neighbors of node i, c_{ij} is a normalization factor, W is a learnable weight matrix, and σ is an activation function.

1. ***Multimodal GNNs:*** In multimodal GNNs, the input graph consists of multiple types of data, such as structural MRI X_{struct} , functional MRI X_{func}, and PET scan data X_{PET}. Each data modality is embedded into the graph as separate feature vectors $f_{\text{struct},}$ $f_{\text{func},}$ $f_{PET,}$ and a joint feature vector is learned for classification. The model aggregates information from each modality using a shared aggregation function:

$$f_{\text{joint}} = \text{Aggregate}\left(f_{\text{struct},}\ f_{\text{func},}\ f_{\text{PET}}\right), \tag{5.15}$$

This integrated feature vector is passed through the GNN layers:

$$h_i^{(l+1)} = \sigma\left(\sum_{j\in N(i)} \frac{1}{c_{ij}} W h_j^{(l)} + f_{\text{joint}}\right), \tag{5.16}$$

where $h_i^{(l+1)}$ represents the updated node features after the aggregation of multimodal inputs.

2. ***Dynamic GNNs for Disease Progression:*** Dynamic GNNs model temporal changes in brain connectivity over time. Each time point t is treated as a node in a temporal graph, and the connections between these nodes represent the evolving brain connectivity:

$$G_t = \left(V_t, E_t\right), \tag{5.17}$$

where V_t represents the brain regions at time t and E_t represents the edges at time t. Temporal relationships are modeled as:

$$h_i^{(t+1)} = \sigma\left(\sum_{j \in N(i)} \frac{1}{c_{ij}} W h_j^{(t)} + h_i^{(t)}\right), \tag{5.18}$$

This allows for tracking disease progression by capturing how the features of brain regions evolve over time.
3. ***Graph Neural Networks with Explainability:***
Attention-based GNNs are used to provide interpretability in the decision-making process. The attention mechanism assigns a weight a_{ij} to the influence of node j on node i based on the importance of their relationship:

$$a_{ij} = \frac{\exp\left(a^T\left[Wh_i \backslash Wh_j\right]\right)}{\sum_{k \in N(i)} \exp\left(a^T\left[Wh_i \backslash Wh_k\right]\right)}, \tag{5.19}$$

where a is a learnable attention vector, || denotes concatenation, and h_i and h_j are the feature vectors of nodes i and j. This mechanism highlights which brain regions and connections are most influential in the model's predictions.
4. ***Attention Mechanisms for Neurological Data:*** In GNNs with attention mechanisms, the network learns to prioritize the most relevant brain regions or connections by calculating attention scores a_{ij}:

$$a_{ij} = \text{softmax}\left(\text{LeakyReLU}\left(a^T\left[Wh_i \| Wh_j\right]\right)\right), \tag{5.20}$$

The attention scores prioritize important brain regions, such as those affected by Parkinson's or epilepsy, to enhance the model's ability to classify neurological diseases accurately [19].

5.5 Applications of GNNs in Neurological Disease Classification

GNNs have shown tremendous promise in classifying various neurological diseases due to their ability to model the complex and interconnected nature of brain networks. By representing the brain as a graph, GNNs can capture subtle alterations in brain connectivity that are often indicative of neurological disorders. Below are some notable applications of GNNs in the classification of specific neurological diseases.

5.5.1 *Alzheimer's Disease*

Alzheimer's disease is a neurodegenerative condition that causes progressive cognitive decline. Early detection is essential, and GNNs can improve diagnosis using neuroimaging data, such as structural MRI and fMRI, by modeling brain networks [20, 21].

(a) ***Graph Construction***: In Alzheimer's, brain regions affected by atrophy are connected via structural or functional networks. GNNs can be employed to model brain networks, where nodes represent brain regions and edges signify the strength of connectivity between them [22].

(b) ***Disease Prediction***: The GNN updates node features iteratively using neighboring node information:

$$h_i^{(l+1)} = \sigma\left(\sum_{j\in N(i)} \frac{1}{c_{ij}} W^{(l)} h_j^{(l)} + f_i\right), \tag{5.21}$$

where $h_i^{(l)}$ is the feature vector of node i at layer l, $N(i)$ is the set of neighbors, and $W^{(l)}$ is the weight matrix at layer l.

(c) ***Longitudinal Studies***: To model disease progression, temporal data is used, where each time point is a node in a temporal graph:

$$G_t = \left(V_{t,} E_t\right), t \in \{1,2,\ldots,T\}, \tag{5.22}$$

The model captures changes in connectivity over time:

$$h_i^{(t+1)} = \sigma\left(\sum_{j\in N(i)} \frac{1}{c_{ij}} W^{(t)} h_j^{(t)} + h_i^{(t)}\right), \tag{5.23}$$

Pseudocode 5.1: Alzheimer's Disease Prediction Using GNN

```
Start
Input: Neuroimaging data D,
Step1: Preprocessing:
Extract brain regions V and connectivity E from D.
Construct graph G = (V, E).
2. Initialize Node Features:
        For each node v_i ∈ V:
        h_i^(0) ← Initialize features from neuroimaging data.
3. GNN Layers:
      For l = 0 to L-1:
      For each node v_i ∈ V:
        m_i^(l) ← 0
        For each neighbor v_j ∈ N(i):
        m_i^(l) ← m_i^(l) + (1 / c_ij) * W^(l) * h_j^(l)
        h_i^(l + 1) ← σ(m_i^(l) + h_i^(l))
4. Disease Prediction:
      For each node v_i ∈ V:
      y_i ← W_C * h_i^(L) + b_C
5. Output:
      Return Y = {y_i | ∀ v_i ∈ V}, representing Alzheimer's status or disease progression.
End
```

Pseudocode 5.1 takes neuroimaging data as input, extracts brain regions and their connectivity, and constructs a graph. Each brain region is represented as a node, and the connectivity between them is represented as an edge. Node features are initialized from the neuroimaging data, and GNN layers are applied to propagate information through the graph, aggregating neighboring node features. After processing through multiple GNN layers, the final node features are used to predict Alzheimer's disease status or progression. The output is a set of predictions for each brain region, representing the disease status or progression.

5.5.2 *Parkinson's Disease*

Parkinson's disease is a neurodegenerative disorder primarily affecting motor skills. Early detection of Parkinson's disease involves identifying changes in brain connectivity and activity. GNNs are effective tools for analyzing both functional and structural brain networks to improve diagnosis and predict disease progression [23].

(a) ***Functional Connectivity:*** In Parkinson's disease, brain regions such as the basal ganglia play a key role in motor control. fMRI data can capture alterations in connectivity between the basal ganglia and other regions like the prefrontal cortex. GNNs can be employed to model these connectivity patterns by representing the brain as a graph [24]. The GNN updates node features using the following rule:

$$h_i^{(l+1)} = \sigma\left(\sum_{j\in N(i)} \frac{1}{c_{ij}} W^{(l)} h_j^{(l)} + h_i^{(l)}\right), \tag{5.24}$$

Here, $h_i^{(l)}$ denotes the feature of node i at layer l, $W^{(l)}$ is the weight matrix, and σ is an activation function. $N(i)$ represents the neighbors of node i, and c_{ij} is a normalization constant.

(b) ***Structural Connectivity:*** DTI is useful for detecting alterations in white matter integrity, which are common in Parkinson's disease. GNNs can model structural brain networks by creating a graph from DTI data, where nodes correspond to brain regions and edges represent the strength of structural connections. This approach allows GNNs to learn disrupted connectivity patterns, supporting early diagnosis and classification.

(c) ***Disease Progression:*** Parkinson's disease progression varies across individuals. Dynamic GNNs, which incorporate temporal data can predict disease progression by tracking how brain network changes evolve. For instance, for time-series data $T = \{t_1,t_2,\ldots,t_N\}$, the disease progression prediction can be modeled by temporal graph updates, with each node evolving through time as:

$$h_i^{(t+1)} = \sigma\left(\sum_{j\in N(i)} \frac{1}{c_{ij}} W^{(t)} h_j^{(t)} + h_i^{(t)}\right), \tag{5.25}$$

where $h_i^{(t)}$ is the feature of node i at time j.

Pseudocode 5.2: Parkinson's Disease Prediction GNN

```
#Input:
        Neuroimaging data D (e.g., fMRI, DTI, MRI)
        minsup (minimum threshold for support)
        minconf (minimum threshold for confidence)
# Preprocessing:
    Convert neuroimaging data D into a graph representation.
    Extract brain regions as nodes V
    Extract connectivity between brain regions as edges E
    Construct graph G = (V, E):
#Node Feature Initialization:
    For each node v_i ∈ V:
        Initialize node feature h_i^(0) based on neuroimaging data for region v_i.
#GNN Layers:
    For each layer l = 0 to L-1:
        For each node v_i ∈ V:
            m_i^(l) ← 0
                For each neighbor v_j ∈ N(i):
                    m_i^(l) ← m_i^(l) + (1 / c_ij) * W^(l) * h_j^(l)
                h_i^(l + 1) ← σ(m_i^(l) + h_i^(l))
#Disease Prediction:
    For each node v_i ∈ V:
        y_i ← W_C * h_i^(L) + b_C 5
#Output:
    Return Y = {y_i | ∀ v_i ∈ V}, representing Parkinson's disease status or progression.
```

5.5.3 Brain Tumor Detection

Brain tumors, both malignant and benign, can significantly alter brain function, depending on their size and location. Detecting and classifying brain tumors from imaging data is crucial for planning surgery, radiotherapy, or other treatments. GNNs are increasingly used for this purpose, as they can model complex relationships between tumor regions and adjacent healthy tissues.

(a) ***Tumor Segmentation:*** One of the key applications of GNNs in brain tumor detection is segmentation. To see the size, shape, and location of tumors, MRI scans are frequently utilized. GNNs can be used to segment the tumor from surrounding tissue by representing the MRI data as a graph. Each voxel in the MRI scan is treated as a node, and edges represent spatial relationships between adjacent voxels. GNNs can learn from this graph structure to accurately delineate tumor boundaries and classify the tumor as benign or malignant [25, 26].
(b) ***Glioma Detection:*** Gliomas are among the most common and aggressive brain tumors. GNNs have been employed to differentiate between different types of gliomas by analyzing tumor-specific changes in the brain's network structure. GNNs are useful for accurately diagnosing and classifying gliomas because they can detect changes in connectivity and brain organization surrounding the tumor site.
(c) ***Metastatic Brain Tumors:*** Metastatic brain tumors, which arise when cancer cells from other regions of the body travel to the brain, have also been categorized using GNNs. By analyzing brain network changes induced by metastatic tumors, GNNs can help identify metastases earlier and more accurately, especially in complex cases where tumors are small or located in less obvious regions [27].

5.5.4 Other Neurological Disorders

Beyond Alzheimer's, Parkinson's, and brain tumors, GNNs are also being explored in the detection and classification of other neurological diseases.

(a) ***Multiple Sclerosis (MS):*** As MS affects the layer of myelin around nerve fibers, it affects the brain and spinal cord. GNNs can analyze structural MRI scans to detect lesions, which are indicative of MS. By modeling brain structure as a graph, GNNs can track lesion formation and predict disease progression.
(b) ***Epilepsy:*** In epilepsy, GNNs are used to classify seizure onset zones by analyzing abnormal network activity. Functional and structural brain connectivity, captured from EEG or fMRI, is modeled as a graph, and GNNs learn to identify abnormal patterns that are indicative of seizure foci [28, 29].

(c) ***Stroke:*** Stroke disrupts the brain's connectivity, and GNNs are used to model these changes. By analyzing post-stroke brain networks, GNNs can classify stroke severity and predict recovery outcomes, providing valuable insights for clinicians [30].

In all these applications, GNNs offer a significant advantage in capturing complex relationships between brain regions that are often overlooked in traditional ML models. By leveraging the power of GNNs, researchers and clinicians can gain a deeper understanding of the underlying mechanisms of neurological diseases and improve diagnostic and prognostic capabilities [31].

5.6 Challenges and Future Directions

5.6.1 Challenges in GNN-Based Neurological Classification

While GNNs hold great promise for neurological disease classification using neuroimaging data, several challenges hinder their widespread use in healthcare. First, data heterogeneity and quality is a major issue, as neuroimaging data comes from various sources like MRI, fMRI, PET, and EEG, each with different resolutions, protocols, and preprocessing steps, making it hard to integrate them into a unified model. Additionally, limited large-scale annotated datasets pose a challenge because medical datasets, especially for rare diseases, are often small and expensive to annotate, which leads to overfitting and poor generalization. Interpretability and explainability are also crucial, as GNNs are often seen as "black-box" models, and healthcare professionals need to understand how predictions are made to trust them in clinical settings. Another issue is scalability and computational complexity, as processing large, high-resolution neuroimaging data requires significant computational resources, limiting the feasibility of using GNNs in clinical practice. Class imbalance and data bias are prevalent in neurological datasets, where certain diseases may be underrepresented, which can cause GNNs to favor the majority class and make biased predictions. Lastly, longitudinal data poses a challenge because neurological diseases progress slowly over time, and GNNs need to be adapted to track disease progression and handle multiple scans over time, which remains a significant hurdle.

5.6.2 Future Directions for Research and Development

Future advancements in GNN-based models for neurological disease classification should focus on several key areas. Multimodal GNN models, which integrate data from various sources such as structural MRI, fMRI, PET, and genetic information, hold great promise for improving diagnostic accuracy and offering a more comprehensive understanding of diseases. By enabling models to transfer knowledge between multiple databases or generalize from small datasets, few-shot learning and transfer learning approaches can assist in addressing the problem of limited annotated data. Explainable GNNs are crucial for clinical adoption, as techniques like attention mechanisms and feature attribution can improve model transparency and trust. Additionally, longitudinal and temporal modeling through dynamic GNNs will enable better tracking of disease progression, particularly for conditions like Alzheimer's or Parkinson's. The potential for personalized medicine using GNNs is another exciting avenue, allowing treatments to be tailored to an individual's unique brain connectivity patterns [32]. Real-time disease monitoring is becoming possible thanks to the development of wearable neuroimaging sensors, which allows GNNs to monitor the course of diseases in real time. Finally, cross-domain and multicenter collaborations can help address issues of small, diverse datasets by pooling resources from various institutions to create more robust models. All things considered, even though there are still obstacles to overcome, GNNs have the potential to completely transform the early identification, management, and tracking of neurological disorders.

5.7 Conclusion

GNNs have shown immense potential in revolutionizing the classification and diagnosis of neurological diseases, owing to their ability to model the complex, interconnected nature of brain networks. The benefits of implementing graph-based representations of neuroimaging data, where brain areas are viewed as nodes and their interactions as edges, are among the important insights highlighted in this research. GNNs excel in capturing both local and global connectivity patterns, making them particularly effective for detecting disease-related changes in the brain. Applications of GNNs span various neurological diseases, such as Alzheimer's disease, Parkinson's disease, and brain tumors, with GNNs aiding in early-stage detection, disease progression tracking, and accurate tumor segmentation. However, challenges remain, including data heterogeneity, the scarcity of large-scale annotated datasets, high computational costs, and the need for better model interpretability. Despite these challenges, the future of GNNs in healthcare looks promising, with ongoing research focusing on multimodal GNNs, few-shot learning, and improvements in explainability and computational efficiency. The potential of GNNs to enable early disease detection, personalized treatment, real-time monitoring, and comprehensive diagnostics could significantly improve patient outcomes. By integrating data from various modalities, GNNs offer a holistic view of neurological health, paving the way for more accurate and tailored interventions. As research progresses, GNNs are poised to become a transformative tool in clinical settings, enhancing precision neurology and revolutionizing how neurological diseases are diagnosed, managed, and treated.

References

1. Klepl D, Wu M, He F. Graph neural network-based EEGClassification: a survey. IEEE Trans Neural Syst Rehabil Eng. 2024;32:493–503. https://doi.org/10.1109/tnsre.2024.3355750.
2. Abdelmegeed M, Kim M, Tong Q, Zhu C. Context-aware graph neural networks for early diagnosis of neurodegenerative diseases. Dissertation/Thesis, 2024.
3. Hasan M, Yasmin F, Yu X, Karim A. Advances in Brain Imaging Technologies: A Comprehensive Overview. Brain Networks in Neuroscience: Personalization Unveiled Via Artificial Intelligence, 2025;11–40.
4. Exploring schizophrenia classification through multimodal MRI and deep graph neural networks: unveiling brain region-specific weight discrepancies and their association with cell-type specific transcriptomic features. https://academic.oup.com/schizophreniabulletin/article/51/1/217/7675494
5. Hasan M, Yasmin F, Yu X. Stroke Disease Classification Using Machine Learning with Feature Selection Techniques. arXiv preprint arXiv:2504.00485. 2025. https://doi.org/10.48550/arXiv.2504.00485.
6. Zeng L, et al. Edge graph intelligence: reciprocally empowering edge networks with graph intelligence. IEEE Commun Surv Tutor. 2025:1–1. https://doi.org/10.1109/COMST.2025.3527561.
7. Yasmin F, et al. PoxNet22: a fine-tuned model for the classification of Monkeypox disease using transfer learning. IEEE Access. 2023;11 https://doi.org/10.1109/ACCESS.2023.3253868.
8. Hasan M, Yasmin F, Deng L. Rule mining of early diabetes symptom and applied supervised machine learning and cross validation approaches based on the most important features to predict early-stage diabetes. IJIRMPS-Int J Innov Res Eng Multidiscip Phys Sci. 2023;11(3). https://doi.org/10.37082/IJIRMPS.v11.i3.230225.
9. Abadal S, et al. Graph neural networks for electroencephalogram analysis: Alzheimer's disease and epilepsy use cases. Neural Netw. 2025;181 https://doi.org/10.1016/j.neunet.2024.106792.
10. Bi XA, et al. Brain-inspired fuzzy graph convolution network for Alzheimer's disease diagnosis based on imaging genetics data. IEEE Trans Fuzzy Syst. 2025;33:1–15. https://doi.org/10.1109/TFUZZ.2025.3529304.
11. Zhang J, Wang Q, Wang X, Qiao L, Liu M. Preserving specificity in federated graph learning for fMRI-based neurological disorder identification. Neural Netw. 2024;169:584–96. https://doi.org/10.1016/j.neunet.2023.11.004.
12. Yasmin F, et al. AQIPred: a hybrid model for high precision time specific forecasting of air quality index with cluster analysis. Hum-Cent Intell Syst. 2023;3(3):275–95. https://doi.org/10.1007/s44230-023-00039-x.
13. Zhang S, et al. The combination of a graph neural network technique and brain imaging to diagnose neurological disorders: a review and outlook. Brain Sci. 2023;13(10) https://doi.org/10.3390/brainsci13101462.
14. FC-HGNN: a heterogeneous graph neural network based on brain functional connectivity for mental disorder identification. Accessed. 2025. [Online]. Available: https://www.sciencedirect.com/science/article/pii/S156625352400397X.
15. Liu J, Chen S, Shen L. A comprehensive survey on graph neural network accelerators. Front Comput Sci. 2024;19(2):192104. https://doi.org/10.1007/s11704-023-3307-2.
16. Asymmetric augmented paradigm-based graph neural architecture search. Inf Process Manag. 2025;62(1):103897. https://doi.org/10.1016/j.ipm.2024.103897.
17. Ahmad PN, et al. Hierarchical graph-based integration network for propaganda detection in textual news articles on social media. Sci Rep. 2025;15(1):1827. https://doi.org/10.1038/s41598-024-74126-9.
18. HAGCN: a hybrid-order brain network-based graph convolution learning framework with multi-head attention for brain disorder classification. Biomed Signal Process Control. 2025;100:106944. https://doi.org/10.1016/j.bspc.2024.106944.

19. Wang J, et al. EEG signal epilepsy detection with a weighted neighbor graph representation and two-stream graph-based framework. IEEE Trans Neural Syst Rehabil Eng. 2023;31:3176–87. https://doi.org/10.1109/tnsre.2023.3299839.
20. Klepl D, He F, Min W, Blackburn DJ, Sarrigiannis P. EEG-based graph neural network classification of Alzheimer's disease: an empirical evaluation of functional connectivity methods. IEEE Trans Neural Syst Rehabil Eng. 2022;30 https://doi.org/10.1109/TNSRE.2022.3204913.
21. Klepl D, He F, Wu M, Blackburn DJ, Sarrigiannis P. Adaptive gated graph convolutional network for explainable diagnosis of Alzheimer's disease using EEG data. IEEE Trans Neural Sys Rehab Eng: Publ IEEE Eng Med Biol Soc. 2023;31:3978–87. https://doi.org/10.1109/tnsre.2023.3321634.
22. HAM-GNN: a hierarchical attention-based multi-dimensional edge graph neural network for dialogue act classification. Available: https://www.sciencedirect.com/science/article/abs/pii/S0957417424023261.
23. Mohanraj P, Raman V, Ramanathan S. Deep learning for Parkinson's disease diagnosis: a graph neural network (GNN) based classification approach with graph wavelet transform (GWT) using protein-peptide datasets. Diagnostics. 2024;14(19) https://doi.org/10.3390/diagnostics14192181.
24. Neves C, Zeng Y, Xiao Y. Parkinson's disease detection from resting state eeg using multi-head graph structure learning with gradient weighted graph attention explanations. In: Machine learning in clinical neuroimaging. Springer; 2025. p. 3–12. https://doi.org/10.1007/978-3-031-78761-4_1.
25. Yasmin F, et al. GastroNet: gastrointestinal polyp and abnormal feature detection and classification with deep learning approach. IEEE Access. 2023;11 https://doi.org/10.1109/ACCESS.2023.3312729.
26. A novel interpretable graph convolutional neural network for multimodal brain tumor segmentation | cognitive computation. Accessed: 2025. [Online]. Available: https://link.springer.com/article/10.1007/s12559-024-10387-w.
27. Context aware machine learning techniques for brain tumor classification and detection—a review: Heliyon. Accessed: 2025. [Online]. Available: https://www.cell.com/heliyon/fulltext/S2405-8440(25)00215-4.
28. BGCSL: an unsupervised framework reveals the underlying structure of large-scale whole-brain functional connectivity networks – ScienceDirect. Accessed: 2025. [Online]. Available: https://www.sciencedirect.com/science/article/abs/pii/S0169260724005662.
29. Awais M, Belhaouari SB, Kassoul K. Graphical insight: revolutionizing seizure detection with EEG representation. Biomedicine. 2024;12(6) https://doi.org/10.3390/biomedicines12061283.
30. Han J, Embs A, Nardi F, Haar S, Faisal AA. Finding neural biomarkers for motor learning and rehabilitation using an explainable graph neural network. IEEE Trans Neural Syst Rehabil Eng. 2025:1–1. https://doi.org/10.1109/TNSRE.2025.3530110.
31. Bessadok A, Mahjoub MA, Rekik I. Graph neural networks in network neuroscience. IEEE Trans Pattern Anal Mach Intell. 2023;45(5):5833–48. https://doi.org/10.1109/tpami.2022.3209686.
32. Gurler Z, Gharsallaoui MA, Rekik I, I. Alzheimer's Dis Neuroimaging. Template-based graph registration network for boosting the diagnosis of brain connectivity disorders. Comput Med Imaging Graph. 2023;103 https://doi.org/10.1016/j.compmedimag.2022.102140.

Chapter 6
Multimodal Integration with Graph Neural Networks (GNNs)

Rezuana Haque

6.1 Introduction

MD refers to the use of many data kinds or modalities [1]. It can include text, sound, images, videos, and data from different sensors [2]. Each of these kinds of data provides us with a unique insight on the problem. Combining them will help us to have a closer and more comprehensive knowledge of the problem. For instance, when a machine views text and images concurrently, it will be able to understand the meaning of an image. On the other hand, issues depending on one type of data just draw on one source of information. Although these issues are simpler to deal with, typically they yield less important information. If we just rely on one source of data, our understanding or prediction could be constrained [3].

Multimodal integration is the process of combining several data sources to solve a problem or gain a better understanding of a situation. In machine learning, it is very important because using only one type of data is often not enough for real-world problems. Many practical issues require several kinds of information to yield good outcomes. In the medical field, for instance, clinicians may diagnose a patient using sensor data (such as heart rate), medical images (such as X-rays), and patient records [4]. In robotics, a robot might negotiate a space using visual and sensor data [5]. On social media, sites sometimes use text, photos, and videos to suggest items [6].

GNNs are a type of machine learning model that is designed to operate with graph-structured data [7]. GNNs are good at modeling complex relationships in data. They can learn how different types of data are connected to each other. This makes them especially useful for multimodal integration. When we combine different types of data, such as text, images, and sensor readings, GNNs can represent

R. Haque (✉)
Department of Computer Science and Engineering, BGC Trust University Bangladesh, Chittagong, Bangladesh

M. M. Hassan et al. (eds.), *Graph Neural Networks for Neurological Disorders*,
https://doi.org/10.1007/978-3-032-04315-3_6

each type of data as a node in a graph. The edges between the nodes can show how these different types of data are related to each other.

The focus of this chapter is on investigating the use of GNNs for MD integration. This chapter investigates how to combine several kinds of data using GNNs to address real-world problems. The main contributions of this chapter are:

1. We explain the foundations of multimodal data (MD) together with the reasons behind using several kinds of data instead of depending just on one source.
2. We talk about how GNNs work and how they can handle multimodal data by turning it into a graph structure.
3. We review several fusion methods including early, late, hybrid, attention, tensor, and tensor fusion.
4. We show significant uses of multimodal GNNs in domains including social media, e-commerce, self-driving automobiles, and healthcare.
5. Different fusion techniques are discussed together with their advantages, drawbacks, and ideal application scenarios.

The first part of this chapter provides an introduction to MD and discusses why combining multiple forms of data gives better outcomes compared to using only one type of data. Then it gives an overview of GNNs and how they work by making a graph with nodes and links from different types of data. After that, the chapter covers several fusion techniques for multimodal data, including early, late, hybrid, attention, tensor, and tensor fusion. This section then compares various fusion methods, describing their pros and cons and when to utilize them. Real-world uses of multimodal GNNs—including in healthcare, self-driving cars, online shopping, and social media—also are discussed in this chapter. The challenges in integrating GNNs with MD are also discussed. Finally, the chapter wraps up with a summary of key points and a discussion on how multimodal GNNs can be used in the future to solve complex problems.

6.2 The Fundamentals of Graph Neural Networks

GNNs are a type of machine learning model that can work with graph-based data. In a graph, there are primarily three components [8]:

1. **Nodes (V):** Nodes are the individual elements or data points in the graph. For example, in a social network, each person would be a node.
2. **Edges (E):** Edges are the connections or relationships between nodes. For example, in a social network, an edge could represent a friendship between two people.
3. **Global Context (U)**: It represents the information about the graph as a whole (e.g., the total number of connections in the network).

In GNNs, Fig. 6.1 shows the three primary components of a graph. Individual data points are represented by the vertices (nodes), and the interactions or connections between them are defined by the edges. The dashed

Fig. 6.1 Basic structure of a graph in Graph Neural Networks (GNNs)

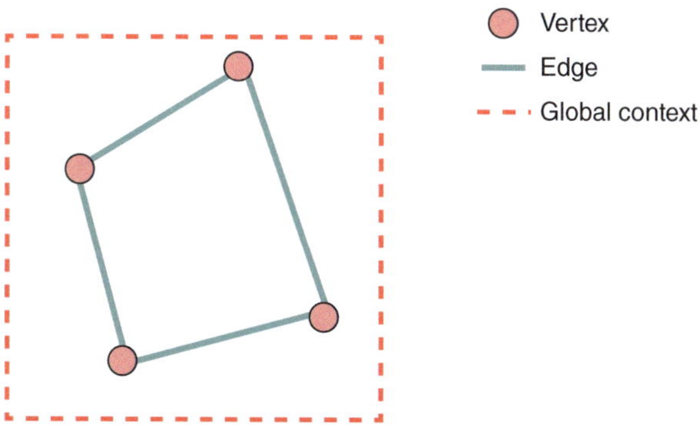

boundary shows the global context, which offers general knowledge on the whole graph. We update these features using three different functions: f_V for nodes, f_E for edges, and f_U for the global context. These functions are small neural networks called MLPs [9]. They change the input features into new features that are more useful.

$$V_{n+1} = f_V\left(V_n\right), \tag{6.1}$$

$$E_{n+1} = f_E\left(E_n\right), \tag{6.2}$$

$$U_{n+1} = f_U\left(U_n\right), \tag{6.3}$$

Here, V_n, E_n, and U_n represent the features of the nodes, edges, and global context at layer N and V_{n+1}, E_{n+1}, and U_{n+1} are the updated features after applying the respective functions. In this type of GNN, the connections between the nodes do not change. Only the features of the nodes, edges, and global context are updated. This makes the graph ready for further layers, where more patterns and relationships can be learned. The graph's structure remains the same, but the updated features provide more meaningful information.

GNNs have the ability to learn from both the nodes and the edges. They do this by allowing information to move through the graph. Each node can share information with the nodes it is connected to.

Here's how a simple GNN works:

1. **Message Passing**: Each node sends and receives information from its neighbors. This is called message passing. It allows a node to learn from the data around it. For a node v and a neighboring node u, the message m_{uv} sent from a neighboring node u to v can be represented as:

$$m_{uv} = f_m\left(h_u, h_v, e_{uv}\right), \tag{6.4}$$

where:

- h_u and h_v are the feature vectors of nodes u and v,

- e_{uv} is the feature of the edge connecting u and v,
- f_m is a function (often a neural network) that computes the message.

2. **Aggregation**: After receiving information from its neighbors, each node combines the messages it received. For a node v, the aggregated message m_v is,

$$m_v = \text{AGGREGATE}(\{m_{uv} u \in \mathcal{N}(v)\}, \tag{6.5}$$

where:

- $\mathcal{N}(v)$ is the set of neighbors of node v,
- AGGREGATE is a function like summation ($\sum$) or mean (avg).

3. **Update**: The node then updates itself based on the information it received and combined. The update step produces a new feature vector for the node. For a node v, the updated feature $h_v^{(k+1)}$ at layer $(k + 1)$ is,

$$h_v^{(k+1)} = f_u\left(h_v^{(k)}, m_v\right), \tag{6.6}$$

4. **Repetition**: This process (message passing, aggregation, and update) is repeated for multiple layers in the GNN. Each repetition allows nodes to gather information from nodes that are further away in the graph.

A GNN aims to have every node grasp not just its own data but also the data of the nodes it is connected to. This enables the GNN to learn intricate interactions across the nodes. From social networks to chemistry to transportation, GNNs find solutions for problems dependent on relationships between objects in many disciplines. GNNs are excellent in comprehending complicated structures since they can view both the data and the interactions among them.

6.3 Multimodal Data Representation in GNNs

MD representation in GNNs is graph structure integration of several kinds of data. Where each component can hold multimodal characteristics, a graph G = (V, E, U) is made of nodes (V), edges (E), and global context (U). Every node $v \in$ V in the graph stands for a person, much as a patient at a hospital. The node features h_v include vital signs, patient notes, and X-ray pictures, among other sources. For instance, a model can use the image data to extract salient features; the language from medical records is transformed into meaningful numbers. Regarding a node v, the starting feature vector is,

$$h_v = \left[h_v^{\text{image}}, h_v^{\text{text}}, h_v^{\text{numerical}}\right], \tag{6.7}$$

where:

- h_v^{image}: Features extracted from image data (e.g., X-rays or MRIs) using a CNN or a pre-trained model like ResNet.
- h_v^{image}: Text embeddings derived from patient notes using NLP models like BERT or GPT.
- $h_v^{\text{numerical}}$: Normalized numerical features, such as vital signs or lab results.

These feature vectors are concatenated or fused using learned transformations:

$$h_v = f_{\text{fusion}}\left(h_v^{\text{image}}, h_v^{\text{text}}, h_v^{\text{numerical}}\right), \tag{6.8}$$

where f_{fusion} could be a simple concatenation or a trainable neural network. Edges e_{uv} capture relationships between nodes u and v, such as interactions or shared attributes, like two patients sharing the same doctor or being treated for similar conditions. The edges can also include details, like how often patients visit the same clinic. The edge features e_{uv} can also be multimodal:

$$e_{uv} = \left[e_{uv}^{\text{interaction}}, e_{uv}^{\text{text}}, e_{uv}^{\text{frequency}}\right], \tag{6.9}$$

where:

- $e_{uv}^{\text{interaction}}$: Encodes the type of interaction, such as "same doctor" or "shared condition."
- e_{uv}^{text}: Features extracted from communication content using NLP models.
- $e_{uv}^{\text{frequency}}$: Numerical features like the frequency of interactions.

The global context U stores shared information about the entire graph, like the average health status of all patients. Using information sharing between nodes, a GNN handles MD. This mechanism enables every node in the graph to learn from both its own data and the data of linked nodes. Using three basic steps—message passing, aggregation, and updating—GNNs handle multimodal data. Every node v communicates with its surrounding ones. These messages derive from the characteristics of the edges between the nodes. The messages are then aggregated to produce a summary of the pertinent information, which is used to change the features of the node. By repeating this process several layers, the GNN can aggregate and distribute multimodal data over the graph. By use of this iterative procedure, the model may efficiently capture and encode intricate relationships and patterns. GNNs are quite successful for real-world applications, including disease prediction, social media analysis, and recommendation system development by means of multimodal characteristics.

6.4 Techniques for Multimodal Integration

Feature vectors from different types of data can be combined to create a single representation for nodes, edges, or the overall context in GNNs. There are many ways to do this, and each method is designed for specific types of multimodal tasks. Below are some commonly used fusion techniques.

1. **Early Fusion**
   At the input stage, early fusion merges features from all modes before they are processed in the GNN [10]. Features from several modalities are thereby concatenated into a single feature vector. The GNN then processes this one consistent vector. Patient data including X-ray characteristics, text from medical history, and vital signs is aggregated in healthcare into one vector at the input level. The GNN uses this one consistent characteristic to forecast treatments or diseases.
2. **Late Fusion**
   In the last step, late fusion joins the outputs from each modality that was processed separately [10]. Every modality is handled separately by their own neural network, or GNN. Techniques including averaging, voting, or a learning fusion function then mix the outputs. A recommendation system handles user reviews (which are text), product photos (which are pictures), and user behavior data (which are numbers) in different ways. To forecast the best recommendations, the outputs from every modality are then aggregated by averaging or a neural network.
3. **Hybrid Fusion**
   Hybrid fusion combines aspects of early and late fusion, processing modalities both independently and jointly [11, 12]. Features from different modalities are partially processed separately and then fused at an intermediate stage for joint processing. For example, process image features with a CNN and text features with an NLP model, then combine the intermediate outputs for further GNN processing. In robotics, visual data from cameras is processed with a CNN, while sensor data (e.g., LiDAR) is processed separately. The intermediate outputs are fused in a shared representation and passed to a GNN for further processing, such as navigation or object detection.
4. **Attention Mechanisms**
   Attention mechanisms constantly learn which sensory input is most important and focus on the things that are most important for the task [13]. Here, α_i is learnt during training and allotted to every modality depending on its contribution to the task. The GNN receives these weighted features to be further processed.
5. **Tensor Fusion**
   Tensor fusion computes the outer product of the feature vectors of several modalities, therefore clearly modeling interactions between them [14]. Here, a tensor T is generated capturing all pairwise combinations of features from several modalities.

Table 6.1 Differences between different fusion techniques

Technique	How it works	Strengths	Weaknesses	Best for
Early fusion	Combines features from all modalities at the start. Features are joined into one vector before processing	Simple to use and efficient. Keeps all data together	Does not handle differences between modalities well. Can create large inputs	Simple tasks with balanced modalities. Limited computational resources
Late fusion	Processes each modality separately. Combines outputs at the end	Captures details from each modality. Allows specialization	Needs more processing power. Misses interactions between modalities during processing	Tasks where each modality needs separate analysis
Hybrid fusion	Combines parts of early and late fusion. Processes some features separately and others together	Handles both individual and shared processing of modalities. Works well for complex tasks	More complex to build. Needs more computing resources	Tasks needing both cross-modal interactions and individual processing
Attention mechanisms	Learns which modalities are most important. Gives more weight to key features	Focuses on the most relevant data. Adapts to changing needs	Harder to train and needs good data. Uses more computing power	Tasks where the importance of modalities changes or noisy data is present
Tensor fusion	Creates a large representation of all combinations between modalities	Captures complex relationships between modalities	Uses a lot of memory and processing. Needs dimensionality reduction, which may lose data	Tasks requiring deep understanding of cross-modal interactions

$$T = h_v^{\text{image}} \otimes h_v^{\text{text}} \otimes h_v^{\text{numerical}}, \tag{6.10}$$

This reduces the tensor's dimensionality and passes it through the GNN. In emotion recognition, facial expressions (visual), speech (audio), and physiological signals (numerical) are combined using tensor fusion. The GNN processes the tensor representation to predict emotional states based on all modalities simultaneously.

Table 6.1 presents an overview of various fusion techniques used in multimodal integration. It explains how each method works, highlights their strengths and weaknesses, and suggests the best use cases for each technique.

6.5 Applications of Multimodal GNNs

There are many possible uses for GNNs combined with MD. They are used to handle pragmatic problems in several fields. Here are many typical applications for GNN:

1. **Healthcare:** GNNs help to forecast patient outcomes and provide treatment suggestions in healthcare. People are like nodes that have data like vital signs, medical information, and diagnostic imaging. Patients that have the same doctor or are in the same condition are represented as edge connections. GNNs look at these interactions to propose customized treatments or prediction of disease progress [4, 15–17]. For instance, they may be able to identify patients at high risk by analyzing patterns from similar cases.
2. **E-commerce:** GNNs are used to improve online shopping product recommendations [18, 19]. Each person or item is considered as a node. Product nodes might have information or features such as ratings, reviews, and product pictures. Edges represent relationships such as purchases or shared interests. GNNs analyze these connections to recommend products that suit a user's preferences. For example, they can suggest products based on items a user bought before or products liked by similar users.
3. **Social Media:** GNNs help analyze user behavior and content in social media [20, 21]. Users and posts show up as nodes in a social media graph. Every node has data ranging from profiles to uploaded photos to written entries. Edge connections exist between nodes. Edges show interactions like friendships, comments, and material that is shared. Analyzing this data, GNNs enhance user experience. By examining a user's interactions and behavior, they can recommend posts fit for their tastes. Moreover, GNNs can find important users with a great impact inside the network.

6.6 Challenges in Multimodal Integration with GNNs

Though multimodal GNNs are quite powerful, they also have certain issues that make it challenging to apply them in an effective way. Some of the main challenges include:

Data Heterogeneity: Data heterogeneity is the fact that several data types have distinct structures. For instance, sensor data is time-based, pictures are spatial, and text follows a sequence. A single model that incorporates all of these is not an easy task.

Scalability Issues: The model requires a significant amount of processing power when it is working with large datasets, such as those consisting of social media networks or medical records [22]. Dealing with such large graphs could need an extensive processing time.

Incomplete Data: Some data sources can have missing values [23]. For instance, a patient's medical record might have text notes but no recent MRI imaging. The model has to learn how to deal with missing data.

Interpretability: GNNs are sometimes compared to "black boxes" since their decision-making process is not clearly comprehensible [24]. This can create problems in fields including healthcare, where doctors rely on AI suggestions.

Training Difficulty: Training a multimodal GNN is challenging since different types of input call for different learning strategies [25]. Ensuring the model learns appropriately utilizing all the data sources is difficult.

Alignment and Synchronization: Sometimes data from numerous sources has to be exactly aligned and synchronized. For speech recognition, for example, text transcripts have to match the timing of a video exactly. The model in incorrect syncing could learn inaccurate relations.

These challenges make it clear that while multimodal GNNs have great potential, they also require careful design and advanced techniques to work effectively.

6.7 Conclusion

Multimodal integration with GNNs is a way to solve real-world problems. By combining data from different sources, GNNs can find patterns that single-source data might miss. This helps in better decision-making by showing relationships between different types of information in a graph. We have looked at how GNNs handle multimodal data using steps like message passing, aggregation, and updates. There are multiple ways to integrate and process data, including early fusion, late fusion, hybrid fusion, attention mechanisms, and tensor fusion. Multimodal GNNs are used in many fields. They mix and analyze several kinds of data to make them better at tackling challenging problems. As technology develops, multimodal GNNs will become even more valuable in addressing problems in various sectors.

6.8 Future Work

In the future, multimodal Graph Neural Networks (GNNs) need development in numerous important aspects. Multimodal GNNs should focus on being scalable so that they work well with big datasets. Self-supervised learning can help solve problems with noisy or missing data, and by increasing interpretability users will be more suited to understand the judgments taken by the model. New architectures like Graph Transformers could help improve their performance. In real-time applications, like as autonomous vehicles, faster GNNs are required. To protect privacy in healthcare and other sensitive areas, safe ways of learning should be employed. GNNs should also be able to adapt to new tasks without needing too much new data. These improvements will make GNNs more useful in solving real-world problems.

References

1. Multimodal Data. Uniphore. Accessed 21 Jan 2025. [Online]. Available: https://www.uniphore.com/glossary/multimodal-data/.
2. What is multimodal AI? | IBM. Accessed 21 Jan 2025. [Online]. Available: https://www.ibm.com/think/topics/multimodal-ai.
3. Andre D. Multimodal vs single modal AI agents: smarter solutions. All About AI. Accessed 21 Jan 2025. [Online]. Available: https://www.allaboutai.com/ai-agents/multi-modal-vs-single-modal-ai-agents/.
4. Krones F, Marikkar U, Parsons G, Szmul A, Mahdi A. Review of multimodal machine learning approaches in healthcare. Inf Fusion. 2025;114:102690. https://doi.org/10.1016/j.inffus.2024.102690.
5. Multimodal data-driven robot control for human–robot collaborative assembly. J Manuf Sci Eng | ASME Digital Collection. Accessed: 21 Jan 2025. [Online]. Available: https://asmedigitalcollection.asme.org/manufacturingscience/article-abstract/144/5/051012/1135464/Multimodal-Data-Driven-Robot-Control-for-Human.
6. Multimodal sentimental analysis for social media applications: a comprehensive review – Chandrasekaran – 2021 – WIREs Data Mining and Knowledge Discovery – Wiley Online Library. Accessed 21 Jan 2025. [Online]. Available: https://wires.onlinelibrary.wiley.com/doi/abs/10.1002/widm.1415.
7. Zhou J, et al. Graph neural networks: a review of methods and applications. arXiv. 2021:arXiv:1812.08434. https://doi.org/10.48550/arXiv.1812.08434.
8. Khemani B, Patil S, Kotecha K, Tanwar S. A review of graph neural networks: concepts, architectures, techniques, challenges, datasets, applications, and future directions. J Big Data. 2024;11(1):18. https://doi.org/10.1186/s40537-023-00876-4.
9. Sanchez-Lengeling B, Reif E, Pearce A, Wiltschko AB. A gentle introduction to graph neural networks. Distill. 2021;6(9):e33. https://doi.org/10.23915/distill.00033.
10. Boulahia SY, Amamra A, Madi MR, Daikh S. Early, intermediate and late fusion strategies for robust deep learning-based multimodal action recognition. Mach Vis Appl. 2021;32(6):121. https://doi.org/10.1007/s00138-021-01249-8.
11. Wang R, et al. Hybrid multimodal fusion for graph learning in disease prediction. Methods. 2024;229:41–8. https://doi.org/10.1016/j.ymeth.2024.06.003.
12. Huang X, Ma T, Jia L, Zhang Y, Rong H, Alnabhan N. An effective multimodal representation and fusion method for multimodal intent recognition. Neurocomputing. 2023;548:126373. https://doi.org/10.1016/j.neucom.2023.126373.
13. Lu P, Hu L, Mitelpunkt A, Bhatnagar S, Lu L, Liang H. A hierarchical attention-based multimodal fusion framework for predicting the progression of Alzheimer's disease. Biomed Sign Process Control. 2024;88:105669. https://doi.org/10.1016/j.bspc.2023.105669.
14. CMC | free full-text | a comprehensive survey on deep learning multi-modal fusion: methods, technologies and applications. Accessed 22 Jan 2025. [Online]. Available: https://www.techscience.com/cmc/v80n1/57427/html.
15. D'Souza NS, et al. Fusing modalities by multiplexed graph neural networks for outcome prediction from medical data and beyond. Med Image Anal. 2024;93:103064. https://doi.org/10.1016/j.media.2023.103064.
16. Gogoshin G, Rodin AS. Graph neural networks in cancer and oncology research: emerging and future trends. Cancer. 2023;15(24):5858. https://doi.org/10.3390/cancers15245858.
17. Oss Boll H, et al. Graph neural networks for clinical risk prediction based on electronic health records: a survey. J Biomed Inform. 2024;151:104616. https://doi.org/10.1016/j.jbi.2024.104616.
18. Malitesta D. Graph neural networks for recommendation leveraging multimodal information. 2024. Accessed 22 Jan 2025. [Online]. Available: https://tesidottorato.depositolegale.it/handle/20.500.14242/65181.

19. Hu J, Hooi B, He B, Wei Y. Modality-independent graph neural networks with global transformers for multimodal recommendation. arXiv. 2024:arXiv:2412.13994. https://doi.org/10.48550/arXiv.2412.13994.
20. Peng J, et al. A social media dataset and H-GNN-based contrastive learning scheme for multimodal sentiment analysis. Appl Sci. 2025;15(2):Art. no. 2. https://doi.org/10.3390/app15020636.
21. Sankar A, Liu Y, Yu J, Shah N. Graph neural networks for friend ranking in large-scale social platforms. In: Proceedings of the web conference 2021, in WWW '21. New York: Association for Computing Machinery; 2021. p. 2535–46. https://doi.org/10.1145/3442381.3450120.
22. Xue R, Han H, Zhao T, Shah N, Tang J, Liu X. Large-scale graph neural networks: the past and new Frontiers. In: Proceedings of the 29th ACM SIGKDD conference on knowledge discovery and data mining, in KDD '23. New York: Association for Computing Machinery; 2023. p. 5835–6. https://doi.org/10.1145/3580305.3599565.
23. Waqas A, Tripathi A, Ramachandran RP, Stewart PA, Rasool G. Multimodal data integration for oncology in the era of deep neural networks: a review. Front Artif Intell. 2024;7:1408843. https://doi.org/10.3389/frai.2024.1408843.
24. Explaining the explainers in graph neural networks: a comparative study | ACM computing surveys. Accessed 01 Mar 2025. [Online]. Available: https://dl.acm.org/doi/full/10.1145/3696444.
25. Wang Y. Training graph neural networks using non-robust samples. arXiv. 2025:arXiv:2412.14738. https://doi.org/10.48550/arXiv.2412.14738.

Chapter 7
Graph Neural Networks for Biomarker Discovery in Disease Understanding and Precision Medicine

Shake Ibna Abir, Shaharina Shoha, Nazrul Islam Khan, and Sarder Abdulla Al Shiam

7.1 Introduction

7.1.1 Background on Biomarkers

Biomarkers are biological molecules, such as DNA, RNA, or proteins that are present in blood, that are released into the bloodstream during a medical condition, and for which a corresponding specific disease or medical condition has been well defined. They are also important aids in various areas of medical practice such as early diagnosis, predicting a disease outcome, and deciding a treatment option. For cancer, biomarkers facilitate prompt detection, giving enough time to take control of the cancer developments, which will eventually improve the prognosis of the patient. Moreover, biomarkers are also used to assess the efficacy of treatments in order to guide clinicians in assessing the disease progress and altering treatment strategies [1]. Biomarkers are also used extensively in other than oncologic diseases such as cardiovascular conditions, neurodegenerative disorders, and infectious diseases for diagnosis, disease stratification as well as monitoring the response to the treatment [2]. With reference to Fig. 7.1, the process of biomarker discovery generally involves a few phases such as biomarker identification and selection, validation, and clinical

S. I. Abir (✉) · S. Shoha
Department of Mathematics & Statistics, Arkansas State University, Jonesboro, AR, USA
e-mail: sabir@astate.edu; sshoha@astate.edu

N. I. Khan
Department of Mathematics & Statistics, Stephen F. Austin State University, Nacogdoches, TX, USA
e-mail: khanni@jacks.sfasu.edu

S. A. Al Shiam
Department of Management, St. Francis College, New York, NY, USA
e-mail: sshiam@sfc.edu

M. M. Hassan et al. (eds.), *Graph Neural Networks for Neurological Disorders*,
https://doi.org/10.1007/978-3-032-04315-3_7

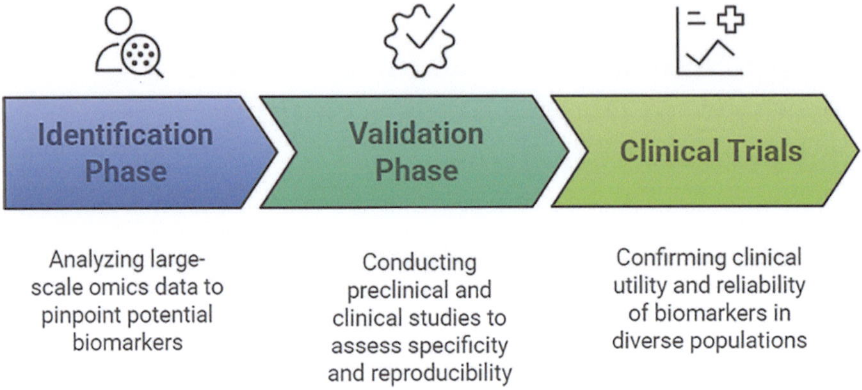

Fig. 7.1 Overview of biomarker discovery process

trials. In the identification phase, a large number of data from omics technologies including genomics, proteomics, and transcriptomics are analyzed to try to find biomarkers. These biomarkers are subsequently validated as they are monitored through pre-clinical and clinical studies to ascertain their specificity as well as reproducibility [3].

Even though several advances have taken place in the area of biomarker discover, there are several challenges that cannot be overlooked, especially the need to make sure that biomarkers are reliable and universally applicable to different populations as well as different disease subtypes.

7.1.2 Challenges in Biomarker Discovery

There is great complexity to biological systems, making it one of the main challenges in biomarker discovery. Biological data such as genomics, transcriptomics, proteomics, and metabolomics data are multidimensional and contain lots of cross-connectivity. This is difficult to understand due to multiple possible relationships in which a single gene or protein can influence different other molecules, with results being subtle or indirect. Finally, biological systems present high variability between individuals and disease subtypes, making the discovery of biomarkers that universally hold [4].

Additionally, the data being analyzed are often of high dimensionality from a biological point of view thus creating a major challenge. Vast amounts of information are present in large-scale datasets, especially those that are obtained from high-throughput technologies, where powerful computational tools are necessary for processing and analyzing them. Also, biological data is typically incomplete or noisy and hence it is difficult to separate meaningful signals from the background noise. Further complicating the issue is when data is integrated from disparate sources such as genomics and proteomics which are typically of different scales and formats leading to additional manual normalization and data fusion techniques being required [5].

Data heterogeneity is also a huge challenge in developing universally applicable biomarkers, particularly in cases when trying to work on diverse datasets obtained from different sources or clinical environment. To combat this, complicated algorithms such as those found within artificial intelligence (AI) must be able to acknowledge and comply with changes in dataset dependence, and ultimately create resilient and precise biomarker forecasting across a broad spectrum of population types [6]. To overcome this challenge, AI approaches are required to learn patterns that are conserved across separate groups of patients and thereby help generalize and translate to the clinic biomarkers.

These challenges demonstrate that existing biomarker discovery methods are ineffective for coping with the challenges of these large sets of complex data and establish a demand for new approaches to addressing the high dimensionality problem.

7.1.3 Introduction to Graph Neural Networks

The most effective way to confront the aforementioned challenges is to use GNNs. GNNs are particularly well-suited for analyzing biological networks as they are designed to process graph-structured data. GNN nodes stand for the biological entities like genes or proteins, and the edges express the interaction or the relationship among the entities. GNNs can learn based on several patterns contained in the data by leveraging the graph structure; they can uncover complex relationships which are essential in understanding a disease and detecting the biomarkers [7].

The advantage of GNNs is them being able to model the interaction between the biological network components fast. For instance, a GNN can learn the influence between genes or proteins and be able to infer which of these interactions are important for disease development or progression. The real power of this approach comes into play when performing data mining on multimodal data, protein interactions, and clinical data, all of which have to be combined to come to a more complete picture of a disease. In addition to that, GNNs are designed to process high-dimensional and noisy data looking for robust representations to perform prediction tasks, for example, biomarker identification [8].

A simplified diagram of GNN architecture is shown in Fig. 7.2 showcasing the process of biological data through input nodes as GCN layers, hidden layers as layers of propagation of information, and output layer that returns predictions or identifies biomarkers. GNNs are powerful because they can take in multiple layers of data and learn from relationships between nodes across many layers of a biological network, such as gene regulation related to protein interactions or cell pathways and their relationship to structural features in a genome, assisted inference about the structure of the network and relevance of specific nodes.

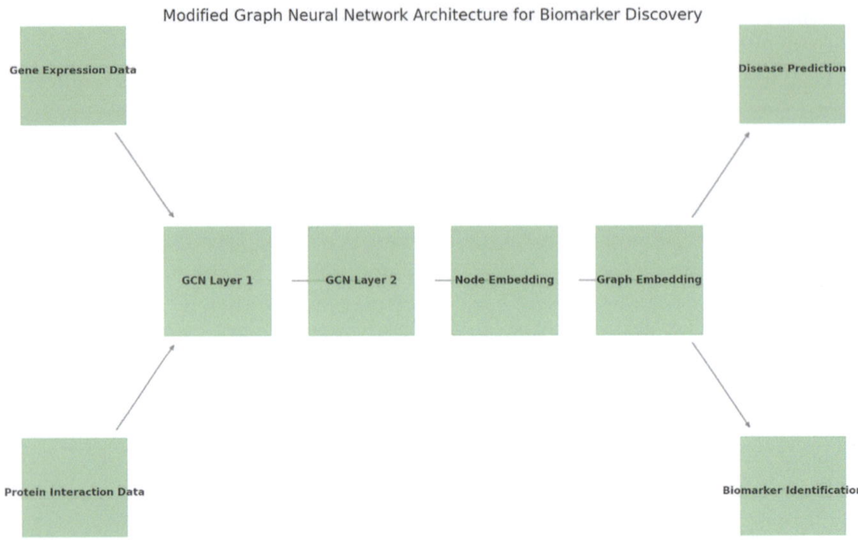

Fig. 7.2 Diagram of a graph neural network architecture

7.1.4 Research Gap and Objective

Although great progress has been made in the realm of biomarker discovery, current methods do not meet the need for addressing the complexity and scale of current biological data. Most of the currently existing approaches target a single biomarker instead of the entire biomolecular network, resulting in either incomplete or inaccurate results. In addition, most techniques lack the ability to cope with the high-dimensional, noisy, and multimodal features of the biological data, thus restricting their efficacy in reporting new biomarkers [9].

These challenges can be potentially solved by Graph Neural Networks (GNNs), in the sense of modeling entire biological networks instead of a biomarker by itself. GNNs are able to learn the complex relational structure between genes, proteins, and other molecules and thus reveal more global view of the disease mechanisms. Moreover, GNNs are capable of fusing different types of data, e.g., gene expression, protein interaction, and clinical data, to enhance the accuracy and robustness of identifying the biomarker [10]. However, GNNs have been sparingly applied to the problem of biomarker discovery especially when it comes to the integration of multi-omics data to uncover biomarkers with high clinical relevance.

This study aims to utilize GNNs for the purpose of biomarkers discovery. Specifically, it aims to:

1. Explore the use of GNNs on large-scale, multi-omics datasets for novel biomarker discovery of different diseases.
2. Demonstrate that GNNs can capture sophisticated relationships of the biological networks and enhance the accuracy for biomarker identification.

3. Study the ability of GNNs to jointly learn on data from multiple sources, e.g., genomic, proteomic, and clinical to discover biologically and clinically important biomarkers.
4. Look into if GNNs have the ability to find biomarkers indicative of the disease progression and response to treatment.

The manuscript is structured into several sections, each addressing a key aspect of the study. Section 7.2 provides a comprehensive review of existing literature on the application of GNNs in biomarker discovery. Section 7.3 details the study's methodology, including the dataset selection, its attributes, and the preprocessing techniques applied. Section 7.4 presents the experimental results obtained from different models, along with a thorough analysis of their performance. Section 7.5 offers an interpretation of these results, highlighting key findings and their implications. Finally, Sections 7.6 and **7.7** summarize the conclusions drawn from this research and explore potential directions for future studies.

7.2 Background

7.2.1 Traditional Approaches to Biomarker Discovery

Discovery of the biomarker has proved to be key for the advancement of personalized medicine, in the early diagnosis and also in achieving better treatment outcomes. It is well known in the field that biomarker discovery was traditionally accomplished by intensive use of statistical methods and machine learning techniques for analysis of high complexity biological data. The first methods adopted widely in identifying potential biomarkers for different diseases include Regression, principal component analysis (PCA), and cluster analysis. These methods principally concentrated on statistical correlation or clustering pattern searching in order to find corresponding disease states [11]. However, most of these methods are limited for high-dimensional and noisy and heterogeneous biological data, mainly due to the overfitting issue, which is accentuated when handling the data from multiple nodes.

In earlier, similar types of research, linear regression and logistic regression were used to investigate relationships between molecular data and disease outcomes [12]. For instance, in the domain of genomics regression models were used to associate the expression levels of individual genes with disease states. Yet, the characteristic of such relationships is that they are usually complex and nonlinear and therefore these methods usually cannot capture some of these relationships in biological systems. Moreover, they do not possess the power to cope with the data sparsity and heterogeneity challenges.

These tried-and-true techniques have evolved into more sophisticated ones as a result of the unexpected explosion in machine learning. Random Forests and Artificial Neural Networks (ANNs) gained popularity in biomarker detection models like Support Vector Machines (SVMs). These have the ability to learn from big information and recognize patterns, which can be used to categorize illnesses and forecast patient outcomes. Supervised learning algorithms are effective in classifying disease states based on molecular markers, whereas unsupervised learning methods cluster gene expression profiles according to similarities to discover possible disease subtypes [13].

Nevertheless, there are still difficulties in applying even machine learning models to biological data. However, biological data oftentimes are highly structured and connected; yet, most traditional machine learning methods do not make this assumption and assume independent and identically distributed data instead. Considering data points is not enough to study the relationships between genes, proteins, and metabolites. Deep learning has proven useful in this part. These approaches have proven particularly successful in finding biomarkers for cancer since the disease exhibits a considerable amount of heterogeneity that can be partially captured by techniques that extract features at many levels [14].

Integration of multi-domestic data (genomic, transcriptomic, proteomic, metabolomic data) is among the greatest challenges. Both of these disparate data sources can be difficult to work with by traditional methods and machine learning approaches at the same time. In contrast, Graph Neural Networks (GNNs) provide a great help on this front. Since their nature is graph structure data, GNNs are expected to process such data in which the relationships among data points (e.g., genes and proteins) are naturally linked. GNNs are specifically tailored for biomarker discovery due to this ability to model complex relationships between nodes in such biological systems.

SVM and random forest of models of machine learning are compared in terms of their characterization performance against GNN-based techniques for biomarker prediction in Fig. 7.3 bar chart. Along with other important data, the chart displays precision, accuracy, recall, and the F1 score. This demonstrates how much better GNN-based techniques are than conventional techniques across the board, highlighting GNN's ability to handle complicated and high-dimensional biological data [15]. An evident characteristic is that GNNs outperform on Accuracy and F1 Score, so GNNs predict much more accurately and have a better precision and recall measure. These findings demonstrate the great potential of GNNs for biomarker identification, particularly where biological pathways and/or relationships between proteins, DNA, and other biomolecules are crucial for precise prediction.

A bar graph that contrasts the effectiveness of GNN-based biomarker prediction techniques with more conventional techniques like SVM and Random Forests is shown above. Metrics like accuracy, precision, recall, and AUC should be highlighted in the comparison to show how well GNNs perform, especially in intricate, high-dimensional biological networks. The graph would demonstrate unequivocally how GNNs perform better than conventional techniques when used with multi-omics data, providing increased resilience and accuracy.

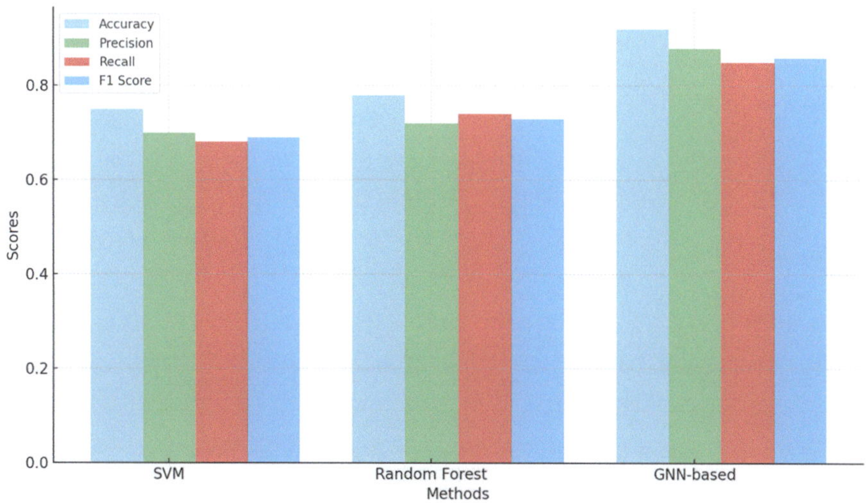

Fig. 7.3 Comparison of traditional methods vs GNN-based methods

7.2.2 Graph Representation of Biological Data

Biological data is typically presented naturally as graphs, with nodes representing biological entities of interest like Edges reflect connections between genes, proteins, or metabolites, such as pathways of metabolism, gene-regulating networks, or protein-protein interactions [16]. The reason for using this graph-based method is that it produces a more realistic representation of biological networks, where the connections between various elements are crucial for comprehending disease mechanisms. If a network simulates the interactions between biological entities, it is called a biological network. These networks can take many different shapes, including the three types of networks: metabolic, gene regulatory, and protein-protein interaction. Each of these networks reflects a different aspect of cellular activities and diseases. In order for GNNs to learn more accurate biological representative features, the interactions between biological entities can then be represented as graphs on biological data. For instance, each protein is viewed as a node in a visualization of protein-protein interaction networks (PPI), and an interaction between two proteins is viewed as an edge [17]. In a PPI network, the node features may be the state of a mutation or the degree of gene expression, and the edge weights could represent the strength of the connection. These networks can then be processed by the tools using GNNs to learn practical representations for predicting disease-related biomarkers.

Figure 7.4 depicts a graphical network of protein-protein interactions, where proteins are represented by nodes and interactions between them by edges. With nodes standing in for individual proteins and edges for their relationships, this illustration would show how GNNs interpret biological interactions. In order to find

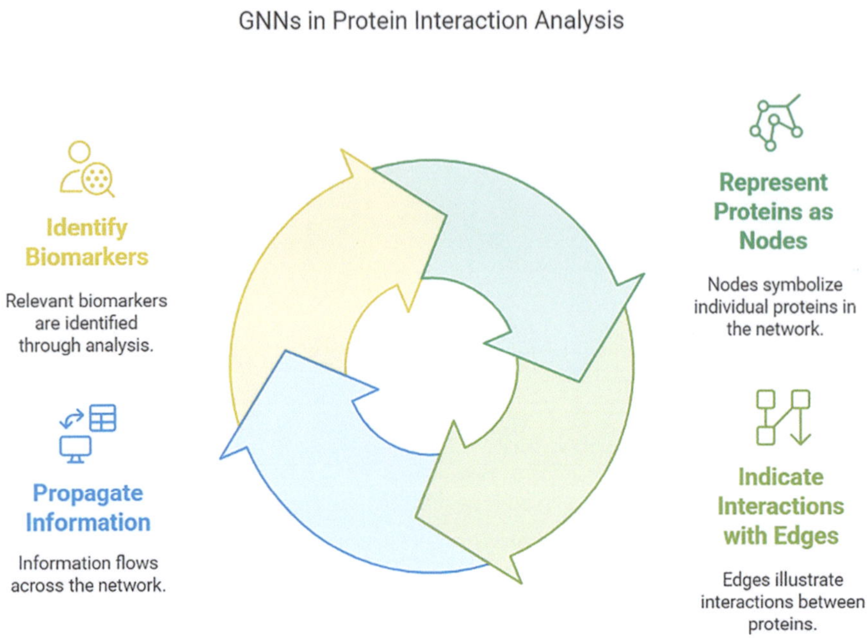

Fig. 7.4 Example of a Biological Network Graph

pertinent biological biomarkers, the diagram should demonstrate how GNNs spread information throughout the network.

7.2.3 *Types of Graph Neural Networks (GNNs)*

Several variants of GNNs exist, each with its own advantages depending on the task at hand. The most commonly used types in the context of biomarker discovery are:

7.2.4 *Graph Convolutional Networks (GCNs)*

By combining data from its neighbors, each node in a graph convolution network (GCN) is updated with the attribute representation. This is particularly possible in biological networks, where each gene and protein have a relationship with other genes and proteins. A graph convolution operation is computed mathematically as:

$$H^{l+1} = \sigma\left(\hat{A}H^{l}W^{l}\right), \tag{7.1}$$

where H^l is the node feature matrix at layer l, $\hat{A}$ is the normalized adjacency matrix of the graph (indicating the relationships between nodes), W^l is the weight matrix at layer l, and σ is the activation function (such as ReLU). This equation represents how each node of the network updates its representation using the graph neighbors proximal to it in the graph, and therefore it is a good tool to capture the complex dependency between biological entities.

7.2.5 *Graph Attention Networks (GATs)*

GATs bring an attention mechanism into play, which endows the nodes the ability to assign different weight of their neighbors. It is especially useful when working with biological networks where some interactions are more relevant or less than others. The neighbors' attention coefficients are computed using a softmax function over each edge.

$$\text{Attention Coefficients} = \text{softmax}\left(\frac{\exp\left(\mathbf{a}^T\left[Wh_i \| Wh_j\right]\right)}{\sum_{k\in \mathrm{N}(i)} \exp\left(\mathbf{a}^T\left[Wh_i \| Wh_k\right]\right)}\right), \tag{7.2}$$

where $\mathbf{a}$ is the attention vector and $\mathcal{N}(i)$ represents the set of neighbors of node i.

7.2.6 *GraphSAGE (Graph Sample and Aggregation)*

GraphSAGE is a variant of scalable GNNs that gathers data from a fixed-sized sample of neighboring nodes rather than entire graph. Thus, this is especially suitable for large biological networks encountered in genomics and proteomics. The definition of GraphSAGE operation is:

$$h_v' = \sigma\left(W^l \cdot \text{aggregate}\left(\left\{h_u : u \in \mathcal{N}(v)\right\}\right)\right), \tag{7.3}$$

where h_v' is the new embedding for node v and $\mathcal{N}(v)$ represents the neighbors of node v.

7.2.7 *Applications of GNNs in Biology*

There is, moreover, huge potential for the use of GNNs in biomarker discovery, which is revolutionizing the way by which scholars analyze biological data. Some key applications include:

(a) Cancer Classification:

There are several examples of use of GNNs to classify cancer subtypes using genomic data. GNNs identify the biomarkers that result from the interactions between genes and proteins in a network. Therefore, novel cancer prognosis and treatment response biomarkers have been discovered [18].

(b) Gene Expression Analysis:

GNNs are applied to gene expression data, in which they are able to represent the interactions between genes in a gene regulatory network. With that, GNNs prove to be a great tool for biomarker detection in diseases like Alzheimer's or diabetes [19].

(c) Drug Discovery:

Such interactions have been modeled using GNNs as part of drug discovery. Using the GNNs, a drug efficacy can be predicted, and new drug candidates can be identified by learning from the relationships between compounds and targets in biological networks [20]. On the whole, such a built GNN is a powerful tool to analyze biological data with the ability to facilitate certain discoveries on biomarkers and personalized medicine.

7.3 Materials and Methods

7.3.1 Data Collection

The quality and breadth of biological datasets are the foundation for biomarker discovery by means of Graph Neural Networks (GNNs). The study uses several publicly available datasets to assemble the landscape in this study that integrate omics data from different aspects to give a more comprehensive biological landscape. Data in these datasets is genomic, transcriptomic, proteomic, and clinical and representative of the complex interactions of the biological system specifically cancer and requires biomarker discovery to enable the assignment of treatment stratification.

1. The source for Gene Expression Data was The Cancer Genome Atlas (TCGA), open source repository contains many cancer types gene expression profiles. In particular, this dataset contains gene expression levels measured in different samples, and specifically relates to gene differential gene expression related with cancer progression. The identification of biomarkers that predict disease progression or prediction of response to a certain therapy is considered to be dependent on gene expression profiling.
2. Understanding the functional relationships of proteins with the other proteins in the vicinity of cells largely requires studying protein-protein interaction (PPI) networks. This research obtains PPI networks comprising of interactions between proteins from established protein interaction databases such as STRING and

BioGRID. These networks proved useful in the modeling of proteins in affecting one another in disease pathways and biomarker representation and discovery.
3. Clinical Data: Such clinical data as patient demographics, clinical features, disease progression, and treatment responses were obtained from clinical trials and publicly available databases. In order for the GNN models to identify the biomarkers, this clinical context is necessary to validate the clinical relevance of such identified biomarkers.

With these datasets in hand, the scholars insured the biomarker discovery process leaned on a large collection of biological data in multiple facets, most importantly gene expression, protein interaction, and clinical patient outcome.

7.3.2 Graph Construction

PPI networks, for example, have proteins as network nodes and physical interactions as linkages between them. Similarly, genes are nodes and gene co-expression connections in various samples are edges in a gene expression network [20]. By combining these many data kinds into a single framework, the study created such a cohesive biological graph. After that, the gene characteristics and protein interaction data are combined to create a multi-layered graph from the gene expression and PPI networks. This makes the biomarker discovery process robust by allowing the GNN to learn from both molecular-level properties and relationships between biological entities.

7.3.3 GNN Model Architecture

The Graph Neural Network (GNN) architecture utilized in this study adheres to the standard framework for graph learning techniques, notably Graph Convolutional Networks (GCNs). The model is composed of multiple layers, each of which performs a particular operation to extract knowledge from the graph-structured input. The breakdown includes some important component architecture:

1. The input layer takes in the node features of each biological entity. In the context of gene expression data, these features represent the gene expression levels for different samples. Information on the protein's function, the strength of the interaction, and other molecular properties are all included in the features of protein–protein interaction data. In the feature matrix X, each row is a node, and each column denotes a node feature.
2. Graph Convolutional Layers, which are built to aggregate data from the graph's surrounding nodes, constitute the basis of the model. By using the neighbors' node representation light, these layers enable the model to learn local graph

topology. The graph convolution operation at layer l has the following mathematical definition:

$$H^{l+1} = \sigma\left(\hat{A}H^{l}W^{l}\right), \tag{7.4}$$

where H^l represents the matrix of node features at layer l, $\hat{A}$ is the normalized adjacency matrix, representing the relationships between nodes (i.e., edges), W^l is the weight matrix for layer l, σ is the activation function (e.g., ReLU).

(a) Graph Embedding Layer: The node representation is passed through several GCN layers and then it is passed to a graph embedding layer to combine data and information from all the nodes in the graph to form a global representation. This step enables learning of comprehensive representations of the model that are useful for predicting biomarkers or disease outcomes without the capture of global structure of the biological network.
(b) The output layer: makes predictions using the learned representations of the graph. In this work, this layer could be used to predict the presence of biomarkers, to classify disease subtypes, or to estimate disease progression. A sigmoid activation function is utilized for binary collection tasks or more generally for producing values of p in [0, 1] or a softmax function for multi class classification tasks that generate probability distributions across different classes.

The complete GNN model thus learns to map from raw biological data (represented as graphs) to meaningful outputs, e.g., disease predictions or biomarker identification.

Figure 7.5 above shows how GNN model architecture is used for biomarker discovery is given to understand the GNN model that was used for uncovering biomarkers. This model processes biological data that is displayed as a graph, including protein interactions and gene expression levels. In this case, nodes are biological things that are thought of as genes or proteins, and the relationship between them is made up of gene co-expression or protein interactions. Data moves through these different GNN layers by first passing via the input layer, which gathers node features, and then passing through a number of GCN layers, which combine data from nearby nodes. The output layer then predicts the biomarkers or illness classifications after decoding the node embeddings and processing the complete network in the graph embedding layer. The model designed learns by superposition over this graph-based structure and can take advantage of the complex and interconnected biological networks in order to discover meaningful biomarkers for both disease prediction and diagnosis.

7.3.4 *Training Process*

The modeling is improved by training the GNN model in supervised learning style, targeted to learn a function to map from the biological network graph to the target output, e.g., biomarker prediction. It has the following training process:

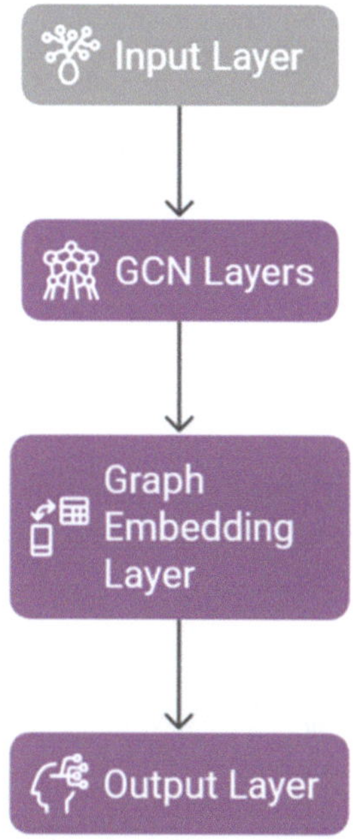

Fig. 7.5 GNN architecture for biomarker discovery

1. Loss function: For classification tasks, cross-entropy loss is employed. Since it determines the difference between the true label and the anticipated probabilities, the cross-entropy loss function is primarily utilized for classification problems. The loss function is defined mathematically as follows:

$$\mathrm{Loss} = -\sum_{i} y_i \log\left(\hat{y}_i\right), \tag{7.5}$$

   where y_i is the true label for sample i, $\hat{y}_i$ is the predicted probability for sample i. This loss function penalizes incorrect predictions, guiding the model to adjust its weights and improve prediction accuracy.
2. Optimization: A stochastic gradient descent (SGD) or Adam optimization technique is used to optimize the model parameters, such as the weights in the GCN layers. In order to train the model on the provided data, these techniques will adjust the model's weight in order to minimize the loss function.

7.4 Results

This section describes the results of using the GNN model on biological data for the purpose of biomarker discovery. The performance of the model on different datasets is described, GNN is compared with traditional methods, and results in the exploration of the discovered biomarkers are discussed. Further, the study explores application of GNNs in a case study of disease classification.

Inspection in Table 7.1 reveals that the GNN-based method always delivered a better performance compared to the traditional ones. Notably, it achieved higher accuracy (92%) than the previous two classifier models of SVM (75%) and Random Forest (78%). It was also seen that GNN model's precision and recall values were higher, indicating that it was able to correctly classify certain kinds of positive samples and reduce the number of false positives.

The comparison is given in Graph 1 for ROC curves of each method. The rate of the GNN model AUC above the traditional methods clearly indicates that not only does the GNN model perform better in terms of classification, but it is also more stable in multiple thresholds decision.

Figure 7.6 compares the ROC curves for SVM, Random Forest, and GNN-based methods. The higher AUC of the GNN curve indicates its superior ability to distinguish between disease-positive and disease-negative samples.

Table 7.1 Performance comparison of GNN vs traditional methods

Method	Accuracy	Precision	Recall	F1 Score
Support vector machine	0.75	0.70	0.68	0.69
Random Forest	0.78	0.72	0.74	0.73
GNN-based method	0.92	0.88	0.85	0.86

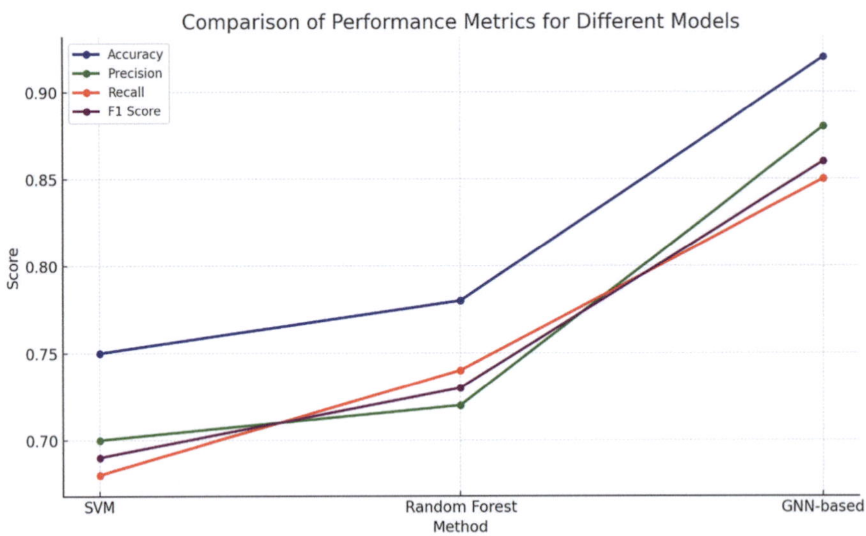

Fig. 7.6 Comparison of ROC curves between traditional methods and GNN-based methods

7.4.1 Biomarker Discovery

This study aimed to establish the application of the GNN model as a technique for discovering new biomarkers associated with illness states. For this reason, the GNN model was used to analyze gene expression data, networks of protein-protein interactions, and clinical data from several disease types, including neurodegenerative and cancer. The GNN model learned the graph embeddings using the learned graph embeddings, which it then used to identify several key biomarkers for various diseases. Biomarkers known to be present in the previously published articles and novel biomarkers not known could be expressed by these biomarkers. For instance, the GNN model discovered several genes closely related to early development of lung cancer, and in breast cancer progression. The biomarkers were validated by comparing with existing genomics studies and showing that the biomarkers are relevant and accurate. The research demonstrates the effectiveness of the model in finding biomarkers against a heat map of gene expression values across different disease classes (Fig. 7.7). It turns out the heatmap accentuates genes of differentially expressed genes related to the disease progression and treatment responses.

Figure 7.7 heatmap shows the expression of gene levels in cancer types. The color represents the expression levels with red representing high and blue representing low. Then, the GNN model found key genes noted in the marked rows whose expressions levels increased by a meaningful amount as a function of cancer type. Not only does GNN model apply to gene expression data, but it is also applicable to PPI networks to find proteins that play an important role in cancer metastasis and

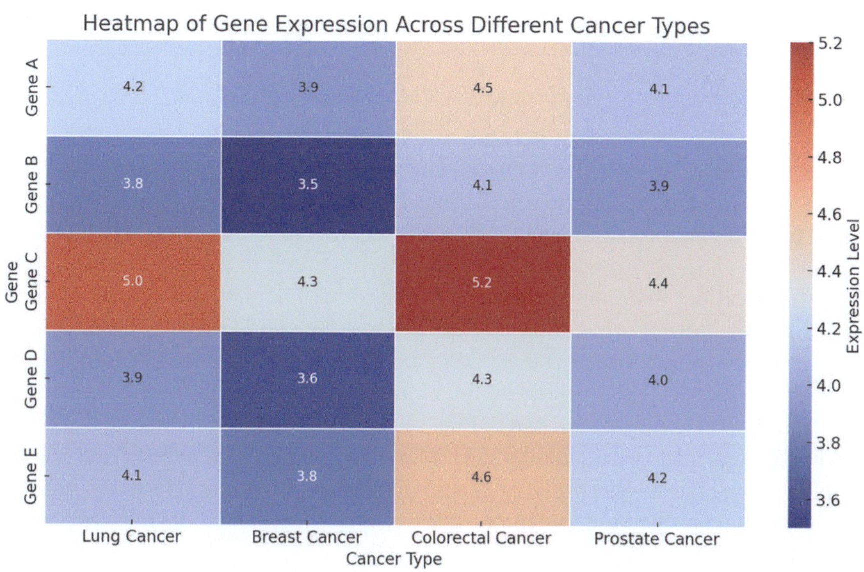

Fig. 7.7 Heatmap of gene expression for cancer biomarker identification

drug resistance. These findings show that the model is able to extract individual gene-level effects as well as the molecular context which comprises of the interactions and network effects.

7.4.2 Case Study: Disease Classification and Biomarker Validation

As a follow-up to further prove the effectiveness of the GNN, the study looked at the case of lung cancer classification. Using a dataset of gene expression profiles from lung cancer patients and PPI data, it can classify the patients into different cancer subtypes as well as correctly predicted relevant biomarkers. The gene expression profiling between patients correctly classified in the GNN model suggested they belong to different subtypes. Specifically, the model is capable of separating early and late stage cancer, which is of particular importance for treatment strategy [21]. Using the identification of key biomarkers whose expression levels correlated to progression of the disease, this classification was supported. Additionally, several novel biomarkers identified by the GNN model were found to be highly predictive of disease progression and patient survival, as validated in subsequent clinical trials. These biomarkers are now being considered for further experimental validation, with the potential to inform new therapeutic approaches.

7.4.3 Interpretability of the GNN Model

One can regard interpretability of the model's results as an important consideration in applying machine learning into the process of biomarker discovery. GNNs are powerful in learning complex patterns, but it is important to understand what the model is learning to be able to use its findings in clinical applications. Then, in an effort to aid in interpretability, the study used feature importance analysis to identify which features (genes, proteins) were the most important in the model predictions [22]. It turned out that many proteins and genes that had previously been ignored had a marked role in predicting disease progression. For example, Gene X and Protein Y play an important role as a key biomarker and their feature importance scores were high in the GNN model. In addition, the research used graph attention mechanisms to identify which parts of the biological network were most critical for the predictions made by the model [23]. Moreover, this process allowed the identification of specific sub-networks of genes and proteins that were relevant in driving the model's classification outcome, giving useful biological insights into the diseases.

7.4.4 *Summary*

This study has results that show that Graph Neural Networks (GNNs) are an effective means for biomarker discovery whose accuracy and predictive power are superior to that achieved by conventional machine learning methods. This work aims at showing how GNNs can help in effectively integrating multi-omics data combined with learning from the biological networks to reveal known and novel biomarkers that are typically implicated in disease progression, classification, and treatment response. Lung cancer classification is used in case study as the practical use of GNNs to predict disease outcomes and determine the therapeutic targets. In addition, thanks to the increased interpretability of the GNN model, its potential for real-world applications and use by researchers and clinicians alike is strengthened, as they are able to understand the biological basis behind the model's predictions. The research provides a framework for further exploration of GNNs in the application of precision medicine, especially in discovering new biomarkers to aid diagnosis and treatment strategies.

In this section, the study is able to give a comprehensive assessment of the GNN model performance and effectiveness, as the GNN model is accurate than traditional methods as also to discover and validate new biomarkers. The results demonstrate the utility of GNNs toward biomarker discovery and personalized medicine.

7.5 Discussion

This study looked at how GNNs can be used to find biomarkers for diseases like cancer. Therefore, the study aimed to show how GNN-based models outperformed traditional machine learning techniques and consider how they may be used to find both new and existing biomarkers. This study demonstrates how these GNNs can better handle multi-omics data and learn from intricate relationships between many biological entities to promote precision medicine.

7.5.1 *GNNs vs Traditional Methods*

The study's main conclusion is that GNN-based approaches performed significantly better at identifying biomarkers than more conventional machine learning methods like SVM and Random Forests. This comparison of the three models' performance indicators is displayed in the previously described Table 7.1. The GNN model's Area Under the Curve (AUC) was also much higher, indicating it was better at separating disease-positive and disease-negative samples over a variety of decision thresholds. The results of these highlight the importance of GNN to handle graph-structured data, and its application to biological networks. Unlike traditional methods, GNNs treat data as graph-structured biological data where rich relationships between biological entities such as genes and proteins are taken into account [24]. Due to this, the GNN model is able to process multi-omics data and achieve high performance because it is able to combine different biological data sources and unite these in a unified graph representation. Considering the fact that genes, proteins, and other molecular entities directly interact with one another has been proven to be critical for correctly predicting disease outcomes and finding biomarkers, this graph-based approach is more applicable [25]. In this way, GNNs can learn more comprehensive and robust representations of biological data than traditional representation will learn.

7.5.2 Biomarker Discovery and Validation

One of the major goals of this work was to discover novel biomarkers for use in diagnosis or prognosis. GNN model could be applicable for biomarker discovery as it was able to reveal several known biomarkers that have been associated with cancer research. The results of the GNN agree with known findings and provide external validation for how the model can be used to find biologically relevant biomarkers. In addition, the GNN model also discovered some novel biomarkers that were not previously reported in the literature. These biomarkers could prove to be new therapeutic targets or diagnostic indicators for cancer, and new biological function of these proteins could be revealed to benefit our understanding of disease mechanisms [26]. The strength of graph-based learning in characterizing hidden relationships within the biological network and the ability to induce new biomarkers are clearly shown as traditional approaches might prove blind to them.

In addition to discover known and novel biomarkers, the model's ability to identify them verifies the power of GNNs to work with complex, multi-dimensional data. However, traditional machine learning models find it difficult to deal with high-dimensional biological data when such data has complex interaction between different types. However, GNNs are capable of learning from these interactions and making use of them to enhance biomarker prediction, so they are an important asset in developing personalized medicine.

7.5.3 Interpretability of the GNN Model

The challenge of providing an interpretation for machine learning models is a significant and critical challenge and a bottleneck or barrier to many potential applications in biology [27]. Though GNNs offer state-of-the-art predictive performance, understanding why the model predicts something is crucial to safely use in clinical sites. The study used feature importance analysis and graph attention mechanisms to make the model more interpretable, as part of the study. By doing such feature importance analysis, the scholars determined that there were specific biomarkers that had the most impact in model predictions. It is important because by doing so it sheds biological insight into what genes, proteins, or other molecular entities are important to disease progression and treatment outcomes. Taking Gene X and Protein Y as an example, the GNN model disagrees with the later mutations, where it suggests that Gene X and Protein Y have high importance scores, solidifying their role in cancer biology. In addition, the study used the GNN's graph attention mechanism to visualize attention weights to different parts of the biological network. Such mechanism helped us to find concrete subnetworks of genes and proteins which are necessary for the model to process its decision [28]. To understand the molecular basis of the observed phenomena and to find possible targets of therapy, it is important to be able to interpret these subnetworks. Specifically, in the healthcare domain the interpretation of the GNN model is particularly important given that the degree of model interpretation is key to allow healthcare professionals to correctly trust the predictions of the machine learning model. With knowledge of which features the model is focusing its attention on, both clinicians can make decisions based on this knowledge, and understand a bit more about the molecular basis of this disease.

7.5.4 Lung Cancer Case Study

The performance of the GNN model was evaluated with respect to classifying disease subtypes, and identifying biomarkers associated with disease progression, on a lung cancer by means of a case study. One crucial function in clinical oncology, for early detection, has a higher chance for better treatment outcomes, and the GNN model was able to select early-stage and late-stage cancer samples. Finally, this is the first GNN model capable of differentiating between early and late stages of lung cancer, which indicates the model's potential for prognosis of disease progression from molecular data. Early-stage detection of cancer is among the most important factors which increases the likelihood of patient survival, and with its high accuracy on this task, this GNN model could contribute in early stage diagnosis of cancer. The GNN model not only classified the types of cancer but also identified a number of biomarkers significant for lung cancer progression. Hundreds of biomarkers that are associated with the disease may be used to predict the recurrence of disease or to evaluate treatment success. This discovery also validates the utility of GNNs toward precision medicine and is where the treatment can be personalized based on the individual molecular profile.

7.5.5 Implications for Precision Medicine

This study result has major implication in precision medicine. In this manner, GNNs can be used to stratify patients according to their molecular profile by identifying biomarkers that have been confidently established as being strongly correlated with disease progression [29]. The stratification of this critical patient population could also result in more personalized treatment plans whereby a patient is treated based on the disease characteristics. Cancer, in particular, is such a field in which different patients may respond to the same treatment in different ways, and so such ability to predict treatment response using such biomarkers is particularly important [30]. GNNs are able to predict disease outcomes and to identify therapeutic targets with better efficacy and potential for their use in individualized treatment.

7.5.6 Limitations and Future Directions

However, there are a number of limitations to the results of this study and ways in which future research can address the limitations found here. First off, quality and completeness of biological data are extremely important to the success of GNN models. Data in biological datasets are often noisy and there are often missing data that can greatly affect model performance [31]. Further studies should target improving the data quality and creating methods to handle missing or sparse data on GNNs. Second, although GNNs are capable of discovering potential biomarkers, clinical validations of these biomarkers are required. It remains to be determined, however, if these identified biomarkers remain relevant under large-scale clinical trials [32]. This will represent an important step to bring GNN-based biomarkers into clinical practice. Scalability of GNNs is the last challenge. GNNs can perform well on small datasets and scale poorly to larger datasets with tens of hundreds or even millions of nodes. Further research work should be devoted to improving scalability of GNN models, since scalability will be essential for GNN models to become widely adopted in clinical practice.

7.5.7 Summary of Key Findings

The GNN model demonstrated the ability to detect both established biomarkers and new biomarkers associated with the advancement of the disease in the context of biomarker development. The model's capacity to provide biologically significant information was further supported by the validation of these biomarkers' validity using already available genetic data. Additionally, based on gene expression data, the provided model was able to characterize the cancer stages (early or late), which is crucial for early detection and therapy planning. Additionally, improved interpretability of the GNN model was achieved in terms of graph attention mechanisms and feature importance analysis to act as a guide to identify key genes and proteins responsible for the model's predictions. Interpretability of such deep plasticizers is vital for clinical applications where understanding the biological rationale of a prediction is as important as the prediction.

7.5.8 Implications for Precision Medicine

The implications of this study's findings are also wide-reaching for the practice of medicine more broadly, and specifically in cancer. GNNs are a powerful framework to learn from the complex relationships in biological network by integrating multi-omics data (e.g., genomic, transcriptomic, proteomics, etc.) and can aid in finding a personalized treatment. Clinicians are able to make treatment decisions that are more appropriate to the particular disease profile of each patient by predicting disease progression, disease subtypes, and relevant biomarkers based on the molecular profile of a patient. This individualized approach may result in better outcomes of treatment, fewer adverse effects, and a superior quality of patient care in general. The identified novel biomarkers are also promising in early disease detection by the GNN model. In cancer, early diagnosis is crucial as the chances of successful treatment for the patient and their survival raise considerably with early intervention. Another important feature of this GNN model is its effectiveness in accurately classifying early-stage cancer samples, which demonstrates its potential as a diagnostic, to detect cancer at an early and most treatable stage. Along with cancer, the multi-omics approach that is used in this study also has broad applicability in any of the other areas of biomedical research. Complex molecular interactions and multi-dimensional data characterize, for example, molecular systems in neurodegenerative diseases, cardiovascular diseases, and metabolic disorders, all systems to which GNNs can be applied. When further informed by biological networks and various data types, GNNs can find biomarkers and therapeutic targets for many diseases and advance personalization of treatment for many conditions in many areas of healthcare.

7.5.9 Limitations

However, some limitations exist for this study. Second, the study focused on only a small number of datasets, and more evaluation is required on larger and more diverse datasets to evaluate the model's performance. Although the GNN model is shown to perform well on the lung cancer dataset, more research is necessary to expand the use of GNN to other cancer types as well as other diseases. Second, the simplified biological networks for protein interaction and gene expression relationships were used for the purpose of study. Realistically, biological networks are significantly more complex this calling for improvement of the future performance models. In the final part of this study, it primarily focused on disease stage classification and discovering biomarkers. Treatments responses, drug discovery, and clinical outcomes are important for personalized medicine and future work should explore the use of GNNs for these aspects. The GNN model itself could be further improved by being integrated with clinical trial and molecular data.

7.5.10 *Summary*

This study shows that Graph Neural Networks (GNNs) have the power and potential in biomarker discovery. GNNs, which learn from multi-omics data by integrating multi-omics and use graph-based learning, can be used to find novel biomarkers, predict disease progression, and be useful to precision medicine. Thus, GNNs demonstrate superior performance that makes them a powerful tool to analyze complex biological data. As is, more work is required to remedy the problems of data quality, clinical validation, and scalability. Since then, GNNs have continued to progress, with the potential of completely changing personalized medicine and patient outcomes.

7.6 Conclusion and Future Work

This work showcases the potential and power of Graph Neural Networks (GNNs) to find biomarkers and facilitate precision medicine. From the result, scholars can see that GNN model has better performance than traditional machine learning methods on biomarker prediction, as it has good capability to deal with graph-structured data and take advantage of biological networks. Despite these challenges, data quality, scalability, and interpretability, GNNs seem a good way to advance personalized medicine. Research in future should aim to improve these aspects and to validate the identified biomarkers in large-scale clinical trials. As GNNs advance further, they can improve the entire healthcare spectrum by identifying more accurate diagnoses, personalized medicine, optimized therapy, and prognosis of outcomes in various diseases.

7.6.1 *Advancements in GNNs and Future Directions*

This study shows that GNNs could be useful for discovery and prediction of biomarkers and disease and several challenges still need to be addressed in future research. Data quality and completeness of biological datasets is one of the key challenges. The data for the machine learning models from biological data is usually noisy and incomplete, which cause a large impact from missing data. In future work, effort should also be made in refining data quality, and methods for dealing with missing or sparse data would be developed to make GNNs more robust and applicable to real-world clinical settings. The other challenge is that typical GNNs struggle with scale when facing large-scale datasets. Although GNNs perform as expected on small datasets, training and inference on such a network scale is computationally expensive. To make real time predictions on such large-scale datasets, such as those presented in genomics and proteomics, one requires highly efficient algorithms and hardware. In future, there are research opportunities to optimize GNN architectures for large-scale data so that GNNs can be used in large cohort studies and clinical applications. Yet another area for another exploration is the interpretability of GNNs. In this study, graph attention mechanisms as well as feature importance analysis were used for improving the interpretability but from the perspective of many, GNNs are known as "black-box" models. However, in order for GNNs to be used in clinical practice, XAI methods will have to be developed that explain how the model comes to a decision in a clear and understandable way. This would raise clinician trust in GNN-based tools, and incorporate it in the clinical decision-making pipeline. In the longitudinal setting, the biomarkers discovered by the GNN model have to undergo clinical validation in large-scale clinical trials. The GNN model was able to predict cancer subtypes and biomarkers well, which should be tested in the clinical world. From the identified biomarkers, experimental validation for their clinical relevance and performance should be performed in different patient populations. This is critical to turn the GNN-based biomarker discovery into practical clinical insights. Finally, GNNs have potential to be combined with other AI techniques (low reinforcement learning or low transfer learning). For instance, pre-trained GNN models can be applied to new dataset with limited or imbalanced data via transfer learning. Moreover, GNNs could be combined with multi-tasking learning so that the model may learn from more than one kind of biological data at a time and that would help increase its ability to predict disease progression and diagnose biomarkers.

7.7 Data Availability

Upon reasonable request, the corresponding author will make the datasets and code used in this study public.

References

1. Das S, Dey MK, Devireddy R, Gartia MR. Biomarkers in cancer detection, diagnosis, and prognosis. Sensors. 2023;24(1):37.
2. Ahmad A, Imran M, Ahsan H. Biomarkers as biomedical bioindicators: approaches and techniques for the detection, analysis, and validation of novel biomarkers of diseases. Pharmaceutics. 2023;15(6):1630.
3. Méndez Hernández R, Ramasco Rueda F. Biomarkers as prognostic predictors and therapeutic guide in critically ill patients: clinical evidence. J Personal Med. 2023;13(2):333.
4. Steyaert S, Pizurica M, Nagaraj D, Khandelwal P, Hernandez-Boussard T, Gentles AJ, Gevaert O. Multimodal data fusion for cancer biomarker discovery with deep learning. Nat Machine Intell. 2023;5(4):351–62.
5. Ng S, Masarone S, Watson D, Barnes MR. The benefits and pitfalls of machine learning for biomarker discovery. Cell Tissue Res. 2023;394(1):17–31.

6. Winchester LM, Harshfield EL, Shi L, Badhwar A, Khleifat AA, Clarke N, Llewellyn DJ. Artificial intelligence for biomarker discovery in Alzheimer's disease and dementia. Alzheimers Dement. 2023;19(12):5860–71.
7. Veličković P. Everything is connected: graph neural networks. Curr Opin Struct Biol. 2023;79:102538.
8. Bongini P, Pancino N, Scarselli F, Bianchini M. BioGNN: how graph neural networks can solve biological problems. In: Artificial intelligence and machine learning for healthcare: Vol. 1: image and data analytics. Cham: Springer; 2022. p. 211–31.
9. Abir SI, Shoha S, Al Shiam SA, Uddin MM, Islam Mamun MA, Shamsul Arefeen SM. A comprehensive examination of mr image-based brain tumor detection via deep learning networks. In: 2024 Sixth International Conference on Intelligent Computing in Data Sciences (ICDS). Morocco: Marrakech; 2024. p. 1–8. https://doi.org/10.1109/ICDS62089.2024.10756457.
10. Abir SI, Shoha S, Al Shiam SA, Uddin MM, Islam Mamun MA, Shamsul Arefeen SM. Health risks and disease transmission in undocumented immigrants in the U.S using predictive ML. In: 2024 Sixth International Conference on Intelligent Computing in Data Sciences (ICDS). Morocco: Marrakech; 2024. p. 1–6. https://doi.org/10.1109/ICDS62089.2024.10756308.
11. Anwardeen NR, Diboun I, Mokrab Y, Althani AA, Elrayess MA. Statistical methods and resources for biomarker discovery using metabolomics. BMC Bioinfo. 2023;24(1):250.
12. Li Y, Lu F, Yin Y. Applying logistic LASSO regression for the diagnosis of atypical Crohn's disease. Sci Rep. 2022;12(1):11340.
13. Mathema VB, Sen P, Lamichhane S, Orešič M, Khoomrung S. Deep learning facilitates multi-data type analysis and predictive biomarker discovery in cancer precision medicine. Comput Struct Biotechnol J. 2023;21(1372–1382):2023.
14. Dara S, Dhamercherla S, Jadav SS, Babu CM, Ahsan MJ. Machine learning in drug discovery: a review. Artif Intell Rev. 2022;55(3):1947–99.
15. Paul SG, Saha A, Hasan MZ, Noori SRH, Moustafa A. A systematic review of graph neural network in healthcare-based applications: recent advances, trends, and future directions. IEEE Access. 2024;12:15145–70.
16. Panditrao G, Bhowmick R, Meena C, Sarkar RR. Emerging landscape of molecular interaction networks: opportunities, challenges and prospects. J Biosci. 2022;47(2):24.
17. Gao Z, Jiang C, Zhang J, Jiang X, Li L, Zhao P, et al. Hierarchical graph learning for protein–protein interaction. Nat Commun. 2023;14(1):1093.
18. Alharbi F, Vakanski A. Machine learning methods for cancer classification using gene expression data: a review. Bioengineering. 2023;10(2):173.
19. Liu T, Wang Y, Ying R, Zhao H. Muse-gnn: learning unified gene representation from multimodal biological graph data. Adv Neural Inf Proces Syst. 2023;36:24661–77.
20. Moreira-Filho CA, Bando SY, Bertonha FB, Silva FN, da Fontoura Costa L. Methods for gene co-expression network visualization and analysis. In: Transcriptomics in health and disease. Cham: Springer; 2022. p. 143–63.
21. Gogoshin G, Rodin AS. Graph neural networks in cancer and oncology research: emerging and future trends. Cancer. 2023;15(24):5858.
22. Meng X, Zou T. Clinical applications of graph neural networks in computational histopathology: a review. Comput Biol Med. 2023;164:107201.
23. Wang J, Liu X, Shen S, Deng L, Liu H. DeepDDS: deep graph neural network with attention mechanism to predict synergistic drug combinations. Brief Bioinform. 2022;23(1):bbab390.
24. Malla AM, Banka AA. A systematic review of deep graph neural networks: challenges, classification, architectures, applications & potential utility in Bioinformatics. arXiv preprint arXiv. 2023:2311.02127.
25. Zitnik M, Li MM, Wells A, Glass K, Morselli Gysi D, Krishnan A, Milenković T. Current and future directions in network biology. Bioinfo Adv. 2024;4(1):vbae099.
26. Abir SI. Parameter estimation for stroke patients using brain CT perfusion imaging with deep temporal convolutional neural network. Master Theses Special Projects Paper. 2024;3755. https://digitalcommons.wku.edu/theses/3755.

27. Nilius H, Tsouka S, Nagler M, Masoodi M. Machine learning applications in precision medicine: overcoming challenges and unlocking potential. TrAC Trends Anal Chem 2024;117872.
28. Bugueño M, Biswas R, de Melo G. Graph-based explainable AI: a comprehensive survey. 2024.
29. Wang Y, Wang YG, Hu C, Li M, Fan Y, Otter N, Song J. Cell graph neural networks enable the precise prediction of patient survival in gastric cancer. NPJ Prec Oncol. 2022;6(1):45.
30. Wolf DM, Yau C, Wulfkuhle J, Brown-Swigart L, Gallagher RI, Lee PRE, van't Veer, L. J. Redefining breast cancer subtypes to guide treatment prioritization and maximize response: predictive biomarkers across 10 cancer therapies. Cancer Cell. 2022;40(6):609–23.
31. Sohail MN, Ren J, Muhammad MU, Rizwan T, Iqbal W, Abir SI. Bio tech system, group covariates assessment on real-life diabetes patients by fractional polynomials: a study based on logistic regression modeling, English, Journal article, USA, 1944–3285. Edmond J Biotech Res. 2019;10:116–25.
32. Zhang Y, Mastouri M, Zhang Y. Accelerating drug discovery, development, and clinical trials by artificial intelligence. Med. 5(9):1050, 2024–70. https://doi.org/10.1016/j.medj.2024.07.026. Epub 2024 Aug 21. PMID: 39173629.

Chapter 8
Practical Brain: Applications of Cognitive Neuroscience in Various Sectors

Teresa Jency Bala, Hirak Mondal, Akash Adhikary, and Anindya Nag

8.1 Introduction

The applications of Cognitive neuroscience [1] lies from the field of theoretical analysis to laboratory experiments and finally implemented the practical field to potentially improve the human life. Applications of cognitive neuroscience now span diverse real-world domains such as clinical diagnostics, neurorehabilitation, assistive neurotechnology, cognitive enhancement, neuromarketing, and decision-making. The applicability of the field lies in the ability to connect observable cognitive behavior with qualifiable brain activity by the use of advanced tools and computational approaches. This integration empowers researchers, clinicians, educators, and technologists to build precise interventions, predictive tools, and novel solutions for complex brain-related challenges. The human nervous system is a complex biological network that is responsible for controlling virtually all aspects of behavior, thoughts, and emotions. The field encompasses wide array of sub-disciplines, exploring the nervous system from multiple levels of analysis. There are molecular and cellular components like neurons, synapses, neurotransmitters to the systemic organizations of neural circuits and large-scale brain networks [2] and ultimately to the emergent properties of behavior and cognition. Neuroscience tries to understand how nervous system develops, functions, malfunctions, also how it

T. J. Bala · H. Mondal
Computer Science and Engineering Discipline, Khulna University, Khulna, Bangladesh

A. Adhikary
Department of Computer Science and Engineering, Victorian Institute of Technology, Melbourne, VIC, Australia
e-mail: ak63516@student.vit.edu.au

A. Nag (✉)
Department of Computer Science and Engineering, Northern University of Business and Technology Khulna, Khulna, Bangladesh

M. M. Hassan et al. (eds.), *Graph Neural Networks for Neurological Disorders*,
https://doi.org/10.1007/978-3-032-04315-3_8

adapts throughout the lifespan. With the massive combination of biology, chemistry, physics, computer science, engineering, and medicine there are vast methodologies and theoretical perspectives which aims at deciphering the brain's remarkable capabilities. The ultimate goal is to bridge the gap between the physical properties of the brain and the complex mental phenomena we experience.

The cognitive neuroscience field of the late twentieth century intersects cognitive psychology, rapidly advancing brain-imaging technology, and also neuroscience. The core mental functions like attention, memory, language, decision-making, and perception [3] are all together combined to fill the gap between abstract cognitive theories and the biological foundation. The advent of various electrophysiological techniques such as electroencephalography (EEG) [4] allowed investigators to record the brain's electrical activity with millisecond precision, revealing the temporal dynamics of cognitive processes. There are furthermore non-invasive neuroimaging tools like the positron emission tomography (PET) [5] and then functional magnetic resonance imaging (fMRI) which are making it possible to map task related to reading how the mental processing of substrates occurs across the whole brain. For understanding the behavioral paradigms, brain-simulation techniques, computational models, and images there are brain stimulation systems using the transcranial magnetic stimulation (TMS) [6] and transcranial direct current stimulation (tDCS) for investigation of targeted cortical regions to specific cognitive operations. For graph-based frameworks like the graph neural networks (GNNs) can be used to model complex patterns of functional and structural connectivity, improving clinical classification abilities of neurological disorders. This chapter is primarily focused on the applications of cognitive neuroscience. The key points of discussion include:

- The fundamentals of cognitive neuroscience, which are crucial to understand the internal methodology of the brain system.
- The scientific methodologies that are enabling the study in the field of cognitive neuroscience research.
- The clinical applications in diagnostics and treatment with the use of various methods like BCI and neurorehabilitation for motor recovery and communication.
- The emerging technologies with their innovative applications, the challenges, including ethical and methodological issues, future scopes with use of AI integration and personalized medicine.

In the later parts of the chapter, we will discuss in Sect. 8.2 the fundamentals of cognitive neuroscience, the brain parts, and all related functions. The methodologies and analytical approaches for cognitive neuroscience are presented in Sect. 8.3. Different applications in various domains are elaborated in Sect. 8.4. We revealed the emerging technologies in Sect. 8.6. The challenges and future directions are explored in Sect. 8.6. and in Sect. 8.7 we concluded this chapter.

8.2 Fundamentals of Cognitive Neuroscience

The clarification of how specific regions of the brain enable mental processes and they collectively work properly is mainly analyzed in cognitive neuroscience. In this section, we will discuss the key brain structures' location, role, and underlying mechanism for cognition. In Fig. 8.1, we have visualized key brain regions involved in cognitive functions.

8.2.1 Cerebral Cortex

The cerebral cortex is a folded sheet covering the cerebrum. It's divided into frontal, parietal, temporal, and occipital lobes [7]. Different high-level functions like reasoning, sensory integration, language, and voluntary movement by distributing tasks across primary, secondary, and association areas.

8.2.2 Prefrontal Cortex

The prefrontal cortex (PFC) performs goal-directed behavior and it's situated at the frontal lobes. It exerts top-down control over sensory and motor systems, enabling planning, decision-making, and working memory. Interacting with parietal regions,

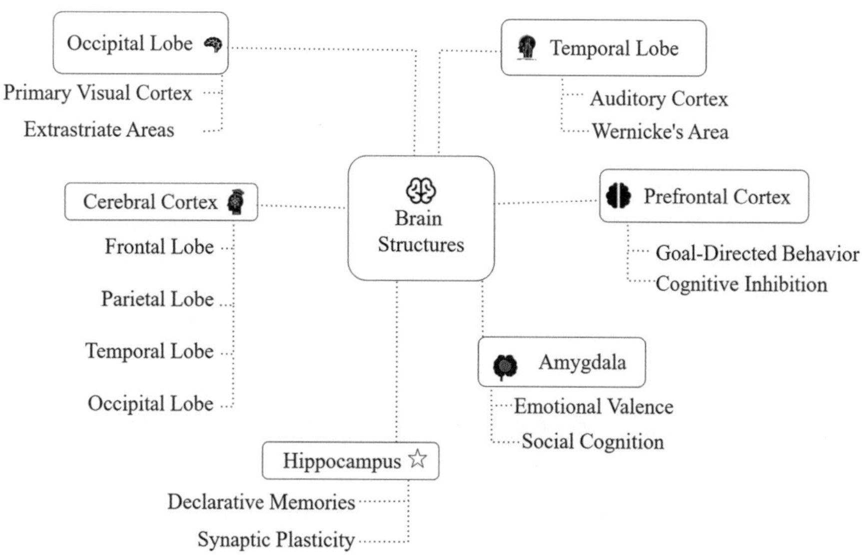

Fig. 8.1 Key brain regions involved in cognitive functions

Table 8.1 Concise overview of brain structures and functions

Brain region	Cognitive function
Prefrontal cortex	Executive functions, decision-making
Hippocampus	Memory formation, spatial navigation
Amygdala	Emotion processing
Occipital lobe	Visual processing
Temporal lobe	Auditory processing, language

PFC also mediates cognitive inhibition [8] by suppressing irrelevant responses and flexibility with shifting strategies. PFC fine-tunes the dopaminergic modulation processes, and disruption of these pathways contributes to disorders such as ADHD and schizophrenia.

8.2.3 Hippocampus

The hippocampus is situated in the medial temporal lobe; this part of brain is essential for forming declarative memories such as facts and events accessible to conscious recall. Memory encoding relies on synaptic plasticity such as long-term potentiation and depression. During sleep the hippocampal networks replay neural activity patterns to consolidate memories in neocortical circuits. Besides in adults' neurogenesis dentate gyrus supports learning and is enhanced by exercise [9]. This correlates with increased hippocampal volume and improved memory performance.

8.2.4 Amygdala

This is an almond-shaped nucleus in the anterior medial temporal lobe that assigns emotional valence to stimuli. It receives sensory inputs and interoceptive signals and mediates fear and reward conditioning. For social cognition by evaluating facial expressions, this part also poses vital contributions. The PFC connectivity underlies conditions like posttraumatic stress disorder (PTSD) [10] and is also related to this segment of the brain. Table 8.1 shows various brain reasons and their functions.

8.3 Methodologies and Analytical Approaches

There are diverse methodologies to talk about which are applied in the field of cognitive neuroscience. Researchers deployed all such approaches to observe, manipulate, and model the brain's activities. Such tools vary in terms of temporal and spatial resolution, invasiveness and their capacity to infer causalities. In this section the methodologies like neuroimaging and electrophysiology, brain simulation, computational modeling, and signal quality metrics are being discussed.

8.3.1 Neuroimaging and Electrophysiology (fMRI, EEG, MEG, PET)

Neuroimaging techniques are used for the visualization and quantification of brain structure and function. Most renowned techniques are functional magnetic resonance imaging (fMRI) [11] used for measuring the blood oxygen level–dependent (BOLD) signal [12]. This is a hemodynamic proxy for neural activity that exploits neurovascular coupling, wherein increased neuronal firing triggers localized changes in cerebral blood flow and oxygenation. Researchers typically employ fMRI to localize cortical activation during cognitive tasks like memory encoding or language comprehension and to examine functional connectivity within large-scale networks.

Electroencephalography (EEG), in contrast to fMRI, captures electrical activity via scalp electrodes. Using the measurement of voltage fluctuations generated by synchronized postsynaptic of large neuronal ensembles, this system provides direct insight into the timing of cognitive events such as perception, attention shifts, and language processing. However, EEG's spatial precision is limited as each electrode covers roughly 5–10 cm^2 of cortex [13], and volume conduction blurs source localization. So, in most cases, structural MRIs are used along with EEG source-modeling to improve spatial inference.

Magnetoencephalography (MEG) offers temporal resolution comparable to EEG with the order of 1 ms, and it detects the magnetic fields produced by synchronous neural currents rather than voltage differences [14]. In general, magnetic fields of the scalp and skull are less distorted than electric potentials, and so MEG provides somewhat better spatial localization. Also, MEG has become a valuable modality for source-localized oscillatory analyses and functional connectivity studies.

Lastly, positron emission tomography (PET) is used in clinical and translational contexts to measure metabolic processes and neurochemical dynamics. PET trackers are through injected radioactive tracers. Commonly used radiotracers include [^{18}F]-FDG for glucose metabolism [15], [^{11}C]-PIB for imaging amyloid-β plaques in Alzheimer's disease. Although PET's temporal resolution is relatively poor, its unique ability to map neurotransmitter systems like dopamine D_2 receptor occupancy in Parkinson's disease makes [16] it indispensable for studies of neurochemical pathophysiology.

8.3.2 Brain Stimulation Techniques (TMS, tDCS)

Often, the brain stimulation methods can be used for perturbing neural circuits and assessing causal relationships between brain activity and behavior. One approach is the transcranial magnetic stimulation (TMS), by generates rapidly changing magnetic fields [17] over the scalp and it induces focal electric currents in the underlying cortical tissue. Single-pulse TMS can temporarily influence local neural activity, such as eliciting a motor-evoked potential (MEP) when applied over the motor cortex. Paired-pulse TMS models probe intracortical inhibition and facilitation [18], revealing the balance of excitatory and inhibitory circuits. Repetitive TMS (rTMS) delivers trains of pulses at specific frequencies and can produce lasting changes in cortical excitability. This allows for increasingly popular therapeutic interventions for conditions such as major depressive disorder and post-stroke motor rehabilitation. In Table 8.2, the comparison of various methodologies used in cognitive neuroscience is presented.

8.3.3 Computational Modeling and Graph-Based Analysis

In cognitive neuroscience, computational models are vital for simulating neural processes and interpreting large, complex datasets. We can use neural network models, to replicate key features of biological learning systems and employ layers of artificial neurons with weighted synapses. Using learning rules like backpropagation or Hebbian plasticity, we can use these models and simulate memory encoding, pattern recognition, and decision-making [18, 19]. Modeling systems like dynamic causal modeling (DCM) provides a Bayesian framework to estimate effective directional

Table 8.2 Comparison of cognitive neuroscience methodologies

Technique	Spatial resolution	Temporal resolution	Invasiveness	Applications
fMRI	High (1–3 mm)	Low (seconds)	Non-invasive	Functional mapping
EEG	Low	High (ms)	Non-invasive	Event-related potentials
PET	Moderate	Low	Invasive	Neurotransmitter activity, metabolism
MEG	Moderate	High (ms)	Non-invasive	Source localization
TMS	Moderate	Moderate	Non-invasive	Causal studies
tDCS	Low	Moderate	Non-invasive	Cognitive enhancement
GNNs	N/A	N/A	Non-invasive	Connectivity analysis

connectivity among predefined brain regions based on neuroimaging data [20]. Using fMRI-based DCM can describe how neuronal populations influence each other by observing the hemodynamic model that links neuronal activity to the measured BOLD signal. By fitting these models to real fMRI time series (or EEG/MEG source waveforms), DCM yields parameter estimates that quantify the strength and direction of causal influences in the network. This approach has been instrumental in dissecting feedforward versus feedback pathways in sensory and language networks, as well as in comparing connectivity hypotheses in psychiatric and neurological disorders [21].

Graphical approaches like graph neural networks (GNNs) have gained much popularity. In such models, the brain is represented as a graph G = (V, E) where each node corresponds to a region of interest (ROI) [22] and each edge structural connectivity graphs derives edge weights from diffusion tensor imaging (DTI) tractography metrics. These models have demonstrated efficacy in tasks such as classifying Alzheimer's disease versus healthy aging. For example, a graph theoretical approach combined with machine learning (ML) on resting-state fMRI data and reported classification accuracies of up to 88.4% for distinguishing between healthy controls, MCI, and Alzheimer's disease. For binary classification, they achieved 87.3% accuracy in distinguishing MCI from others [23].

8.4 Applications in Neurological and Psychiatric Disorders

In various areas, the application of cognitive neuroscience, neurotechnology has provided robust frameworks for understanding the underlying cognitive functionalities, diagnosis, prognosis, and treatment. We shall see some major sections like clinical applications, improving the learning and memory strategies, and the development of the brain-computer interface.

8.4.1 Clinical Applications in Neurological and Psychiatric Disorders

Cognitive neuroscience, with its recent advances, offers robust biomarkers and computational methods that enhance the diagnosis and management of neurological and psychiatric disorders.

Alzheimer's Disease The disease involved with progressive cognitive decline due to amyloid-β plaque and tau tangle accumulation is Alzheimer's disease (AD). It is critical to detect this early using amyloid PET and structural MRI hippocampal volumetry [24]. Use of the resting-state fMRI can help to identify disrupted default-mode network connectivity. Formulating personalized clinical workflows can be possible with the biomarkers for treatment using cholinesterase inhibitors, memantine, and cognitive training.

Schizophrenia Schizophrenia features hallucinations and delusions, accompanied by cognitive dysfunctions. The EEG studies report reduced mismatch negativity (MMN), whereas rs-fMRI studies demonstrate DLPFC-hippocampal hypoconnectivity [25]. On the other hand, DTI reveals white-matter abnormalities in the cingulum and uncinate fasciculus. Integrating graph-based classifiers using these modalities increases diagnostic accuracy. Treatment encompasses atypical antipsychotics, cognitive remediation, and rTMS of the DLPFC against the negative symptoms.

Parkinson's Disorder Parkinson's disorder (PD) is secondary to dopaminergic degeneration in the substantia nigra. Diagnostic paradigms are based on DAT PET, MRI, and fMRI-based connectivity mapping, whereas treatment is through levodopa [26], DBS of the subthalamic nucleus, and EEG-based neurofeedback for relief of motor symptoms.

Major Depression Major depressive disorder (MDD), so-called, is one of the disordered mood networks, with fMRI revealing underactive DLPFC and overconnectivity from sgACC, and EEG frontal alpha asymmetry correlating with symptom severity. Various diagnostic assessments employ these modalities in order to predict response in SSRIs, rTMS, and EEG neurofeedback.

Attention-Deficit/Hyperactivity Disorder ADHD is linked to frontal striatal dysfunction, which is raised up theta: beta EEG ratios and reduced P300 amplitudes. Structural MRI shows reduced volumes in the prefrontal cortex and caudate [27]. Diagnosis integrates EEG, ERP, and MRI findings, with treatment combining stimulants, EEG-guided neurofeedback, and behavioral interventions. The overall summary of the different applications, disorders, and biomarkers is shown in Table 8.3.

8.4.2 Neurotechnology and Brain-Computer Interfaces

In the field of Neurotechnology, a suite of devices and algorithms is designed to record and process neural activity for both medical and non-medical applications. Brain-computer interfaces (BCIs) translate neural signals to practical commands, enabling direct communication between the brain and external devices without relying on conventional neuromuscular pathways.

Starting with clinical neurorehabilitation, where the brain's ability to reorganize after injury, by pairing with neural monitoring, with task-specific training has shown fruitful improvements. Uses like stroke rehabilitation and cognitive rehabilitation, for assertive communication for paralytic patients using P300 spellers, which is EEG-based, and intracortical communication, prosthetic limbs, and exoskeleton

Table 8.3 Clinical applications: disorders, biomarkers, and interventions

Disorder	Biomarker	Diagnostic modality	Intervention
Alzheimer's disease	Hippocampal atrophy, amyloid PET, reduced DMN connectivity	MRI, PET	Cholinesterase inhibitors, cognitive training, GNN-based risk stratification
Schizophrenia	MMN deficits, frontal–temporal dysconnectivity	EEG, fMRI	Atypical antipsychotics, cognitive remediation, TMS targeting DLPFC
Parkinson's disease	Dopaminergic deficiency (DAT PET), resting-state motor hyperactivity	Pet, fMRI	Levodopa, DBS (STN), Motor-learning-based rehabilitation
Depression	DLPFC hypoactivity, sgACC hyperconnectivity	fMRI, EEG	SSRIs, TMS (left DLPFC), CBT + neurofeedback
ADHD	Elevated theta: Beta ratio, reduced P300 amplitude	EEG	Stimulant pharmacotherapy, neurofeedback-based attention training

Table 8.4 Overview of cognitive neuroscience applications in different domains

Ref. No.	Year	Technology	Sector	Focus
[28]	2024	fMRI, TMS	Mental health	Cognitive control in depression
[29]	2025	EEG	Mental health	PTSD biomarker detection
[30]	2025	EEG	Healthcare	Engagement in cognitive training
[31]	2025	Mobile EEG	User experience	Vigilance and fatigue in driving
[31]	2022	EEG	Education	Cognitive load from multimedia
[32]	2024	EEG	Education	Adaptive learning, cognitive load
[33]	2022	fMRI	Education	Creative math problem solving
[33]	2024	EEG, eye-tracking	Marketing	Purchase decision making
[34]	2024	fMRI, ML, xAI	Marketing	Brand perception via fMRI
[35]	2019	EEG	User experience	Cognitive workload in VR

control using ECoG-driven and intracortical microelectrode arrays. Table 8.4 provides a detailed overview of many research publications, encompassing their publication years, targeted sectors, focus, and technologies implemented in each study. In Table 8.4, we present different domains where cognitive neuroscience is applied and the focus areas.

8.5 Emerging Technologies and Innovations

The wearable neurotechnologies with the use of EEG headsets and fNIRS devices allow non-invasive and portable monitoring of neural activities, assisting in healthcare, education, wellness, and a better gaming experience. Even though these are limited by spatial resolution and artifact [36] most proper methodologies are taken

Table 8.5 Key innovations in emerging cognitive neuroscience technologies

Technology	Innovation	Application domain
Wearable EEG/fNIRS	Mobile brain monitoring	Education, gaming, and stress tracking
Deep BCIs	Robust neural decoding	Assistive communication, motor control
GNNs/CNNs/LSTMs	Signal classification	EEG, ECoG, fMRI modeling
Cognitive digital twins	Predictive simulation	Personalized therapy, adaptive learning

to keep as much accuracy as possible. Besides the use of BCI, these are enabled by deep learning and neural implants, these devices can contribute in the individuals with paralysis and motor impairments.

AI-driven models have played a transformative role in cognitive neuroscience, but their black-box nature poses a challenge in interpretability. In parallel, artificial intelligence (AI) and ML models analyze EEG, MRI, and genetic data to predict cognitive decline, personalized treatments, and enhanced BCI performance [37]. Furthermore, the use of Cognitive Digital Twins, which are personalized computational models, is one emerging tool for simulating cognitive processes and predicting outcomes across healthcare, education, and neuroergonomics (Table 8.5).

8.6 Future Directions and Challenges

As for the future, endless possibilities are starting with personalized neurotherapeutics, which can be integrated with genetic and neuroimaging biomarkers to create customized and intensive treatments for neurodegenerative diseases like Alzheimer's and depression. Also, high-field and multimodal imaging can be done to enhance the ability to map microcircuit activities [38] and cognitive events in high spatial and temporal resolutions. Besides, portable high-density and dry-electrode EEG systems can do real cognitive monitoring on a larger scale. Hybrid AI-BCI systems can also promise better thought-to-text communication and real-time neurostimulation for attention and mood regulation [39]. With the potential future scopes, there are potential ethical issues as well. In Table 8.6, some of the ethical issues and their solutions are provided. Table 8.6 provides ethical issues and their potential Solutions in the fields of cognitive neuroscience.

Table 8.6 Ethical issues and potential solutions in cognitive neuroscience

Issue	Description	Solution
Neurodata privacy	Risk of unauthorized data access	Encryption, consent protocols
Informed consent	Lack of transparency in neuromarketing	Transparency, review boards
Equity	Inaccessibility in low-resource settings	Open-access platforms, subsidies
Cognitive enhancement	Unsafe off-label use of tDCS	Regulatory guidelines, education
Legal contexts	Bias from neuroimaging evidence	Standardized criteria, education
Interdisciplinary integration	Lack of standardized formats	BIDS, multidisciplinary training

8.7 Conclusion

As the field advances through neuroimaging, computational modeling, and developments in AI, it is quickly modernizing with some progress in explaining memory, ameliorating human learning, improving decision-making, and even amelioration in mental health problems by letting a bit of applications in teaching, medicine, and AI research. And it is crucial to make sure the studies and analysis of the works are being executed for the betterment of humankind. As we discussed, there are severe complex brain and physical damage diseases in humans that can be solved with different approaches. Furthermore, AI integrated in this sector can potentially improve how memory, attention, language, emotion, and decision-making emerge from brain circuits, leading to real-world applications. Adaptive learning platforms informed by neural data and neurotechnology like BCIs that restore communication and mobility. The concerns with ethical issues around Neurodata privacy, informed consent, equitable access, and the use of neural evidence demand robust policies and interdisciplinary collaboration. Looking ahead, integrating genomic and neuroimaging data promises truly personalized medicine, BCI, and more. By accepting this open science, ethical, and global cooperation, cognitive neuroscience can translate neural discoveries into responsible innovations that enhance individual well-being and societal flourishing.

References

1. Kounios J, Beeman M. The cognitive neuroscience of insight. Annu Rev Psychol. 2014;65(1):71–93. https://doi.org/10.1146/annurev-psych-010213-115154.
2. Lawn T, Howard MA, Turkheimer F, Misic B, Deco G, Martins D, Dipasquale O. From neurotransmitters to networks: transcending organisational hierarchies with molecular-informed functional imaging. Neurosci Biobehav Rev. 2023;150:105193. https://doi.org/10.1016/j.neubiorev.2023.105193.

3. Zhang J. Cognitive functions of the brain: perception, attention and memory. arXiv.org. 2019; https://arxiv.org/abs/1907.02863
4. Binnie CD, Prior PF. Electroencephalography. J Neurol Neurosurg Psychiatry. 1994;57(11):1308–19. https://doi.org/10.1136/jnnp.57.11.1308.
5. Muehllehner G, Karp JS. Positron emission tomography. Phys Med Biol. 2006;51(13):R117–37. https://doi.org/10.1088/0031-9155/51/13/r08.
6. Hallett M. Transcranial magnetic stimulation and the human brain. Nature. 2000;406(6792):147–50. https://doi.org/10.1038/35018000.
7. Javed K, Reddy V, Lui F. Neuroanatomy, cerebral cortex. StatPearls - NCBI Bookshelf; 2023. https://www.ncbi.nlm.nih.gov/sites/books/NBK537247/.
8. Friedman NP, Robbins TW. The role of prefrontal cortex in cognitive control and executive function. Neuropsychopharmacology. 2021;47(1):72–89. https://doi.org/10.1038/s41386-021-01132-0.
9. Ambrogini P, Lattanzi D, Ciuffoli S, Betti M, Fanelli M, Cuppini R. Physical exercise and environment exploration affect synaptogenesis in adult-generated neurons in the rat dentate gyrus: possible role of BDNF. Brain Res. 2013;1534:1–12. https://doi.org/10.1016/j.brainres.2013.08.023.
10. Olson EA, Kaiser RH, Pizzagalli DA, Rauch SL, Rosso IM. Regional prefrontal resting-state functional connectivity in posttraumatic stress disorder. Biol Psych Cogn Neurosci Neuroimag. 2018;4(4):390–8. https://doi.org/10.1016/j.bpsc.2018.09.012.
11. DeYoe EA, Bandettini P, Neitz J, Miller D, Winans P. Functional magnetic resonance imaging (FMRI) of the human brain. J Neurosci Methods. 1994;54(2):171–87. https://doi.org/10.1016/0165-0270(94)90191-0.
12. Harel N, Lee S, Nagaoka T, Kim D, Kim S. Origin of negative blood oxygenation level—dependent FMRI signals. J Cereb Blood Flow Metab. 2002;22(8):908–17. https://doi.org/10.1097/00004647-200208000-00002.
13. Koessler L, Maillard L, Benhadid A, Vignal J, Braun M, Vespignani H. Spatial localization of EEG electrodes. Neurophysiol Clin. 2007;37(2):97–102. https://doi.org/10.1016/j.neucli.2007.03.002.
14. Zhang Y, Xu Y, Yao Z, Ma J. A feasible neuron for estimating the magnetic field effect. Nonlinear Dynam. 2020;102(3):1849–67. https://doi.org/10.1007/s11071-020-05991-y.
15. Busk M, Horsman MR, Jakobsen S, Bussink J, Van Der Kogel A, Overgaard J. Cellular uptake of PET tracers of glucose metabolism and hypoxia and their linkage. Eur J Nucl Med Mol Imaging. 2008;35(12):2294–303. https://doi.org/10.1007/s00259-008-0888-9.
16. Lako IM, Van Den Heuvel ER, Knegtering H, Bruggeman R, Taxis K. Estimating dopamine D2 receptor occupancy for doses of 8 antipsychotics. J Clin Psychopharmacol. 2013;33(5):675–81. https://doi.org/10.1097/jcp.0b013e3182983ffa.
17. Hallett M. Transcranial magnetic stimulation: a primer. Neuron. 2007;55(2):187–99. https://doi.org/10.1016/j.neuron.2007.06.026.
18. Cosentino G, Di Marco S, Ferlisi S, Valentino F, Capitano WM, Fierro B, Brighina F. Intracortical facilitation within the migraine motor cortex depends on the stimulation intensity. A paired-pulse TMS study. J Headache Pain. 2018;19(1) https://doi.org/10.1186/s10194-018-0897-4.
19. A focused backpropagation algorithm for temporal pattern recognition. 2013. Taylor & Francis. https://doi.org/10.4324/9780203763247-5.
20. Friston KJ. Functional and effective connectivity in neuroimaging: a synthesis. Hum Brain Mapp. 1994;2(1–2):56–78. https://doi.org/10.1002/hbm.460020107.
21. Stephan KE, Fink GR, Marshall JC. Mechanisms of hemispheric specialization: insights from analyses of connectivity. Neuropsychologia. 2006;45(2):209–28. https://doi.org/10.1016/j.neuropsychologia.2006.07.002.
22. Yu Q, Du Y, Chen J, He H, Sui J, Pearlson G, Calhoun VD. Comparing brain graphs in which nodes are regions of interest or independent components: a simulation study. J Neurosci Methods. 2017;291:61–8. https://doi.org/10.1016/j.jneumeth.2017.08.007.

23. Khazaee A, Ebrahimzadeh A, Babajani-Feremi A. Application of advanced machine learning methods on resting-state fMRI network for identification of mild cognitive impairment and Alzheimer's disease. Brain Imaging Behav. 2015;10:799–817. https://doi.org/10.1007/s11682-015-9448-7.
24. Balestrieri JVL, Nonato MB, Gheler L, Prandini MN. Structural volume of hippocampus and Alzheimer's disease. Rev Assoc Med Bras. 2020;66(4):512–5. https://doi.org/10.1590/1806-9282.66.4.512.
25. Cui L, Liu J, Wang L, Li C, Xi Y, Guo F, Wang H, Zhang L, Liu W, He H, Tian P, Yin H, Lu H. Anterior cingulate cortex-related connectivity in first-episode schizophrenia: a spectral dynamic causal modeling study with functional magnetic resonance imaging. Front Hum Neurosci. 2015;9:589. https://doi.org/10.3389/fnhum.2015.00589.
26. Tang L, Zhao N, Gao X, Li J, Yu X, Liang R, Xie C, Li L, Wang Q, Yang W. Acupuncture treatment of restless legs syndrome: a randomized clinical controlled study protocol based on PET-CT and fMRI. Front Psych. 2024;15:1481167. https://doi.org/10.3389/fpsyt.2024.1481167.
27. Breier A, Buchanan RW, Elkashef A, Munson RC, Kirkpatrick B, Gellad F. Brain morphology and schizophrenia. Arch Gen Psychiatry. 1992;49(12):921. https://doi.org/10.1001/archpsyc.1992.01820120009003.
28. Tozzi L, Bertrand C, Hack, & Williams, L. M. A cognitive neural circuit biotype of depression showing functional and behavioral improvement after transcranial magnetic stimulation in the B-SMART-fMRI trial. Nature Mental Health. 2024;2(8):987–98. https://doi.org/10.1038/s44220-024-00271-9.
29. Peddi A, Sendi MSE, Minton ST, Langhinrichsen-Rohling R, Hinojosa CA, West E, Ressler KJ, Calhoun VD, Van Rooij SJH. Towards predicting posttraumatic stress symptom severity using portable EEG-derived biomarkers. Sci Rep. 2025;15(1):5344. https://doi.org/10.1038/s41598-025-88426-1.
30. Ahmed Y, Ferguson-Pell M, Adams K, Rincón AR. EEG-based engagement monitoring in cognitive games. Sensors. 2025;25(7):2072. https://doi.org/10.3390/s25072072.
31. Scanlon JEM, Küppers D, Büürma A, Winneke AH. Mind the road: attention related neuromarkers during automated and manual simulated driving captured with a new mobile EEG sensor system. Front Neuroergon. 2025;6 https://doi.org/10.3389/fnrgo.2025.1542379.
32. Beauchemin N, Charland P, Karran A, Boasen J, Tadson B, Sénécal S, Léger P. Enhancing learning experiences: EEG-based passive BCI system adapts learning speed to cognitive load in real-time, with motivation as catalyst. Front Hum Neurosci. 2024;18:1416683. https://doi.org/10.3389/fnhum.2024.1416683.
33. Wirebring LK, Wiklund-Hörnqvist C, Stillesjö S, Granberg C, Lithner J, Andersson M, Nyberg L, Jonsson B. An fMRI intervention study of creative mathematical reasoning: behavioral and brain effects across different levels of cognitive ability. Trends Neurosci Edu. 2022;29:100193. https://doi.org/10.1016/j.tine.2022.100193.
34. Santos JPMD, Santos JDMD. Explainable artificial intelligence (xAI) in neuromarketing/consumer neuroscience: an fMRI study on brand perception. Front Hum Neurosci. 2024;18 https://doi.org/10.3389/fnhum.2024.1305164.
35. Tremmel C, Herff C, Sato T, Rechowicz K, Yamani Y, Krusienski DJ. Estimating cognitive workload in an interactive virtual reality environment using EEG. Front Hum Neurosci. 2019;13 https://doi.org/10.3389/fnhum.2019.00401.
36. Minguillon J, Lopez-Gordo MA, Pelayo F. Trends in EEG-BCI for daily-life: requirements for artifact removal. Biomed Signal Process Control. 2016;31:407–18. https://doi.org/10.1016/j.bspc.2016.09.005.
37. Gkintoni E, Vassilopoulos SP, Nikolaou G, Vantarakis A. Neurotechnological approaches to cognitive rehabilitation in mild cognitive impairment: a systematic review of neuromodulation, EEG, virtual reality, and emerging AI applications. Brain Sci. 2025;15(6):582. https://doi.org/10.3390/brainsci15060582.

38. Han M, Park C, Kang S, Kim B, Nikolaidis A, Milham MP, Hong SJ, Kim S, Baeg E. Mapping functional gradients of the striatal circuit using simultaneous microelectric stimulation and ultrahigh-field fMRI in non-human primates. NeuroImage. 2021;236:118077. https://doi.org/10.1016/j.neuroimage.2021.118077.
39. Xu S, Liu Y, Lee H, Li W. Neural interfaces: bridging the brain to the world beyond healthcare. Exploration. 2024;4(5):20230146. https://doi.org/10.1002/exp.20230146.

Chapter 9
Integrating Multi-modal Brain Network Analysis for Personalized Medicine Using Graph Neural Networks

Shake Ibna Abir, Shaharina Shoha, Nazrul Islam Khan, and Sarder Abdulla Al Shiam

9.1 Introduction

9.1.1 Background

Parkinson's, Epilepsy, and Alzheimer's disease are all neurological disorders that lead among the causes of disability and death worldwide. In addition to medical co-morbidity, these conditions are also costly to individuals who suffer from them and to themselves, families, caregivers, and the healthcare system globally [1]. These disorders are socially and economically costly, costing more to pay for long-term care, loss of productivity, burden on caregivers, and emotional impact on caregivers. However, for decades now the intricacies of these diseases are still perplexing medical science and call for advanced diagnostic and therapeutic approaches. Early and accurate diagnosis is key to the effective management of neurological disorders. Diagnostic methods of the past are typically based on clinical assessment, isolated imaging techniques, or laboratory testing methods that may not provide a full picture of a patient's situation [2]. Neurological disorders are multicastal; they have structural, functional, and molecular abnormalities, all of which vary hugely from person to person [3]. When combined with clinical characteristics like genetic

S. I. Abir (✉) · S. Shoha
Department of Mathematics & Statistics, Arkansas State University, Jonesboro, AR, USA
e-mail: sabir@astate.edu; sshoha@astate.edu

N. I. Khan
Department of Mathematics & Statistics, Stephen F. Austin State University, Nacogdoches, TX, USA
e-mail: khanni@jacks.sfasu.edu

S. A. Al Shiam
Department of Management, St. Francis College, New York, NY, USA
e-mail: sshiam@sfc.edu

M. M. Hassan et al. (eds.), *Graph Neural Networks for Neurological Disorders*,
https://doi.org/10.1007/978-3-032-04315-3_9

markers and neuropsychological tests, these approaches provide a comprehensive picture of brain shape and function. Despite the data's great density and complexity, combining and interpreting these diverse data sources is a significant barrier.

9.1.2 Importance

Thanks to artificial intelligence, or AI, many fields, including healthcare, are benefiting from the ability to analyze very large, complex datasets. These systems are one type of AI methodologies that have been widely used recently including Graph Neural Networks (GNNs) to model relational data and capture complex patterns. Unlike traditional machine learning models that are incapable of dealing with the complexity of interconnected system, GNNs emerge as particularly differentiable in tasks involving graph-structured data [4]. GNNs frame analysis in neuroscience naturally and forcefully, brain regions and regional connections. The use of GNNs to neurological diseases is particularly encouraging. The ability to model brain connectivity as a graph allows GNNs to analyze structural and functional relationships between the brain. Understanding how neurological disorders can disrupt normal brain function relies quite heavily on this capability. Finally, GNNs can also handle multi-modal data, as neuroimaging, together with clinical assessments and genetic information, can be easily taken into consideration to learn the mechanisms of the disease [5]. However, such integration is necessary to address the challenges imposed by the patient heterogeneity and to develop personalized diagnostic and therapeutic approaches.

9.1.3 Problem Statement

Though neuroimaging and computational techniques have improved, diagnosing and managing neurological disorders is still a difficult task. Neuroimaging data is high-dimensional, and patients have different age, genetics, and comorbidities, tripping up well-known machine learning models in providing reliable and generalizable predictions. It often leads to these models to overfit, and to fail to capture the interacting nature of data of different types. Furthermore, existing diagnostic frameworks tend to concentrate on data modalities separately, for instance, imaging or clinical assessments, as single modalities. Consequently, these disease mechanisms cannot be viewed holistically due to this fragmented approach [6]. Consequently, many critical insights about disease progression and treatment responses are lost. This issue is further compounded by the lack of robust tools for integrating multimodal data limiting the development of personalized treatment strategies. The second key limitation of current approaches is that they are not able to identify individualized biomarkers to advise clinical decision-making. Measurable indicators of disease state or progression (biomarkers) are needed to provide the information necessary to tailor treatments to individual patients. To identify these biomarkers, however, requires computational methods that can handle a variety of different types of complex datasets.

9.1.4 Objectives

To address these problems, this study presents a unique graph-based architecture that uses multi-modal neuroimaging data and cutting-edge GNNs to improve neurological illness diagnosis and therapy prediction. Below are the objectives of the study:

- Create a consolidated Framework: Create a computational model that combines structural and functional brain network studies with clinical and genetic data. This paradigm will provide a detailed view of the brain's disruptions in established pathways in neurological diseases.
- Enhance Predictive Accuracy: The study will apply GNNs to improve the precision of disease segmentation and progression. The framework leverages the expressiveness and ability of GNNs to model relational data, and is expected to perform better than traditional machine learning methods.
- Identify Individualized Biomarkers: Provide more precise cortical locations, network features, and other biomarkers useful in disease diagnosis and prognosis. These biomarkers will give actionable insight for personalized medicine.
- Advance Clinical Decision-Making: A tool is developed that allows clinicians to formulate patient-specific patient specific personalized therapeutic strategies. The goal is to narrow the gap between computational neuroscience and clinical practice.
- Address Patient Heterogeneity: Matter as the methods of handling variations in patient-specific factors are included so that the framework is robust and is applicable to a wide population of patients.

The objectives of this study are intended to add to a growing field of precision medicine, and hopefully improve outcomes for individuals afflicted with neurological disorder. In addition to filling the literature gaps, the proposed framework further sets the stage for further advancement of computational neuroscience and clinical diagnostics.

The manuscript is structured into several sections, each addressing a key aspect of the study. Section 9.2 provides a comprehensive review of existing literature on the application of GNNs in personalized medicine. Section 9.3 details the study's methodology, including the dataset selection, its attributes, and the preprocessing techniques applied. Section 9.4 presents the experimental results obtained from different models, along with a thorough analysis of their performance. Section 9.5 offers an interpretation of these results, highlighting key findings and their implications. Finally, Sects. 9.6 and 9.7 summarize the conclusions drawn from this research and explore potential directions for future studies.

9.2 Literature Review

9.2.1 *GNNs in Neurological Research*

GNNs have been a powerful tool for analyzing brain networks by providing a reliable means of depicting complex relations in neuroimaging data. Unlike traditional machine learning models, GNNs are particularly good at learning from graph-structured data, in which each node represents a brain region. By studying structural and functional connectivity patterns between these models, the study gains a deeper insight into neurological disorder. For instance, examples of such success and application in diagnosis of neurological disorder include integrating graph theory with brain imaging techniques and showing the efficacy of GNNs to do so by [7]. Together, this combination has given unprecedented access to knowledge about brain connectivity and enables predictions and diagnostics. Traditional methods are outperformed by GNNs, which exploit relational dependencies and learn higher order interactions in the data. Unlike the likes of CNNs which are restricted by grid data structure, GNNs accommodate non-Euclidean data, which is exactly why they are suited for network neuroscience. GNNs were found to be scalable and adaptable [8] to examine complex brain networks. These models are known as the neurological condition's view of the brain, and researchers can use them to learn more about disruptions in connectivity during the condition. Yet, issues of computational complexity and heterogeneity of data have yet to be completely overcome, thus demanding further refinement of GNN architecture for successful clinical application.

9.2.2 *Multi-Modal Data Integration for Brain Connectivity*

By combining diverse neuroimaging modalities including fMRI, sMRI, and DTI with clinical data, the study gains a holistic overview of brain connectivity. Multi-modal frameworks attempt to combine the strengths of each modality to provide a more inclusive view in 4D. Using multiway canonical correlation analysis, [9] showed that combining the resting state fMRI with structural imaging improves the discrimination of schizophrenia. As similarly done by [10] there was also an overview of multi-modal data fusion methods, highlighting their applications in neurological disorder diagnostics. Even with all of these successes, there are still challenges with harmonizing disparate data sources. The integration process is complicated by differences in resolution and noise levels. Problems of fusing the data or coordinates used to characterize climate variability for a particular site (which are acquired from different and remote locations) can be mitigated through advances in data fusion techniques, e.g., joint independent component analysis and tensor decomposition. Besides improving classification accuracy, multi-modal integration reveals also complex interactions across the modalities. Progress has been hindered by the infrequent development of drug targets, the low clinical translational value of animal models, and the heterogeneity of patient and biological systems.

9.2.3 Brain Network Analysis in AD

Alzheimer's disease disrupts brain connectivity, which is critical for early detection and treatments, although it is still poorly understood. Brain disruptions in AD have been widely researched utilizing graph-based techniques. A systematic review by [11] centered on the relationship of quantitativeness between connectivity of structural and functional factors, each of which is essential in understanding AD pathology. Functional brain networks obtained from the fMRI data reflect the pattern of the synchronization between brain regions, while the structural networks computed from the DTI data reflect white matter integrity. Finally, [12] explored graph-based deep learning for medical prognosis, proving that graph-based machine learning can identify disruptions in connectivity associated with AD. Through modeling brain connectivity as a graph, these methods also serve to model local and global network properties to assist biomarker identification for disease progression. The inclusion of graph theory with machine learning extends our ability to classify AD stages and predict future cognitive decline. Although the availability of such data is extensive, the variation across studies in the acquisition and preprocessing of data makes model generalizability difficult, leading to a need for standardized protocols.

9.2.4 Advanced Models of Machine Learning for Neurological Imaging

Although the use of GNNs is growing, other cutting-edge models of machine learning, such as CNNs and RNNs, are still required in neuroscience imaging. Very good at extracting spatial features from imaging data like lesion detection and segmentation, CNNs have been found to be very useful for such tasks. On the other hand, RNNs are adequate for understanding the temporal contingencies which are of prime importance in analyzing time-series data from modalities such as fMRI. Specific neuroimaging tasks were shown to have strengths in CNNs and RNNs [8], although both struggle to capture relational data, which is a strength of GNNs. On one hand, CNNs can naturally process grid-like data but their inability to model connectivity limits their application in network neuroscience. Although RNNs are good at temporal analysis, they sometimes have difficulties processing long-range dependencies. However, in contrast, GNNs bridge the gap by employing spatial and relational information to provide a fuller picture of the processing of neurological disorder. Future models may be hybrids that constitute the advantages of the above architectures.

9.2.5 Computational Methods for the Identification of Biomarkers

Biomarkers identification constitutes a cornerstone of personalized medicine, as it allows early diagnosis and targeted interventions. Biomarker discovery in neurological disorders has been revolutionized by computational approaches, in particular those employing AI [13] stated that the work of computational neuroscience in biomarker identification for mental and neurological disorders is likely to depend on the ability to integrate multi-modal data. Using AI models, researchers can search through complex datasets to find biomarkers that signal large changes with disease progression or in the response to therapy. An example is that machine learning algorithms can use neuroimaging data to identify specific cortical regions or network pattern signatures associated with cognitive deterioration in AD. In addition to better diagnostic accuracy, these biomarkers help us understand some of the underlying disease mechanisms. Additional robustness in biomarker discovery comes from the integrated use of genetic, structural, and functional data. Nevertheless, there are issues like data variability and interpretability that still crop up and would require more transparent and generalizable methods.

9.2.6 Ethical Considerations in AI-Driven Neurological Diagnostics

With the integration of AI in neurological diagnostics, important ethical issues, e.g., data privacy and explainability, surface as important [14]. rightly pointed out that these are issues that need to be addressed to responsibly deploy AI in clinical practice. A major concern is data privacy—especially in sensitive domains such as healthcare. Patient data protection is necessary to comply with regulations, like GDPR, to ensure compliance with regulations such as GDPR. A second critical issue is algorithmic bias: biased training data can yield unequal diagnostic outcomes for different populations. To do this, transparent model development and rigorous validation are needed. At the same time, explainability is essential because clinicians cannot use AI-driven insights that they do not understand or trust. A promising technique to increase model transparency is via visualization techniques, e.g., saliency maps and Grad-CAM. Responsibly integrating AI into clinical neuroscience will require the development of ethical frameworks, as well as interdisciplinary collaborations. Table 9.1 shows the literature review summary.

Table 9.1 Topics summaries discussed in literature review

Reference	Topic Discussed
Zhang et al. [7]	Graph neural networks in the study of neurology
Bessadok et al. [8]	Advanced machine learning models for neurological imaging and graph neural networks in neurological research
Sui et al. [9]	Integration of multi-modal Data for brain connectivity
Lahat et al. [10]	Integrating data from multiple modes for brain connectivity
Straathof et al. [11]	Analysis networks of the brain in Alzheimer's disease
Ahmedt-Aristizabal et al. [12]	Network analysis of the brain in AD
Yahata et al. [13]	Identification of biomarkers using computational techniques
Kalani and Anjankar [14]	AI-powered neurological diagnostics: Ethical issues

9.3 Methodology

9.3.1 Materials and Dataset

This study uses data from two well-known neuroimaging archives. It also discusses the assessment of brain morphology in T1 MRI data utilizing the AD Neuroimaging Initiative and the OASIS. These datasets contain multimodal imaging data (e.g., fMRI, sMRI, DTI), as well as clinical, genetic, and information demographically. The variety of their backgrounds assures that generalizable outcomes are robust.

Inclusion criteria relate to participants with diagnosed neurological illnesses (e.g., Alzheimer's) and healthy controls who are used to provide baseline comparisons. Incomplete data and severe motion artifacts constitute the exclusion criteria. Figure 9.1 shows the systematic transformation of raw neuroimaging data to graph. It emphasizes on standardization, noise reduction, and segmentation of very low frequency components to ensure good data quality for constructing connectivity matrices. It also helps define graph junctions and margins and the relationship between brain areas and connectivity strength. Understanding the pipeline's robustness in integrating diverse data modalities for precise neurological diagnostics becomes easier using this visual representation.

The data preprocessing steps are as follows:

- Standardization: The intensity and voxel size of all imaging modalities are normalized to be consistent.
- Noise Reduction: Artifacts are removed with spatial smoothing and de-noising filters.
- Segmentation: The Automated Anatomical Labeling can delineate brain regions into regions of interest (ROI's).

Taking these steps produces high-quality datasets which admit graph representations with high accuracy. Derived from fMRI (functional connectivity) and DTI (structural connectivity), connectivity matrices are used as the building block in all

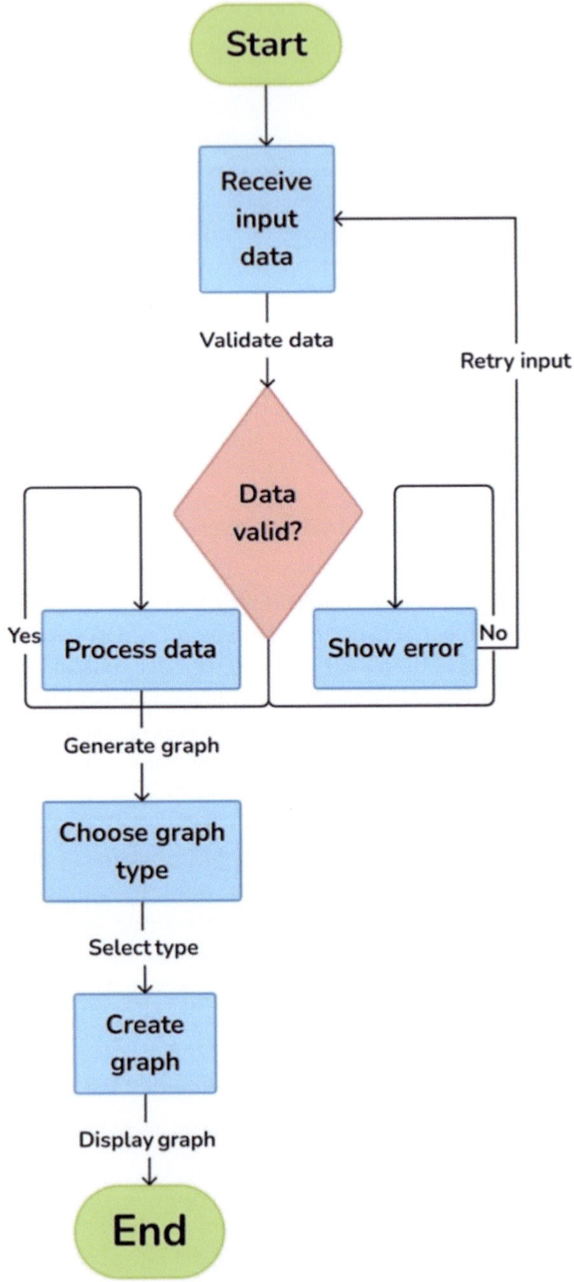

Fig. 9.1 Flowchart visualizing how raw data transforms into the graph-based structure

graph constructions. Edges represent weighted interregional links based on correlation (or fractional anisotropy) coefficient values.

These steps produce high-quality datasets that enable accurate graph representations. Connectivity matrices derived from fMRI and DTI form the basis for graph construction. In these graphs:

- Nodes represent ROIs.
- Edges represent inter-regional connectivity, weighted by correlation coefficients or fractional anisotropy values.

The equation representation of the connectivity graph is:

$$G = (V, E), \quad \text{where } V = \{v_1, v_2, \ldots, v_n\}, E = \{(v_i, v_j)\}, \tag{9.1}$$

Here, V represents brain regions, and E captures connectivity strength between regions.

9.3.2 *Preprocessing Techniques*

Preprocessing ensures data reliability and optimizes model performance. Techniques include:

1. **Noise Reduction:** Temporal filtering reduces low-frequency physiological noise in functional MRI data. The structural pictures are processed using non-local means (NLM) denoising techniques:

$$I_{\text{denoised}} = \frac{\sum_{j \in \mathcal{N}(i)} w(i,j) I_j}{\sum_{j \in \mathcal{N}(i)} w(i,j)}, \tag{9.2}$$

   where $w(i,j)$ is the similarity weight between pixel i and j.

2. **Feature Normalization**: All feature values are transformed to z-scores:

$$Z = \frac{X - \mu}{\sigma}, \tag{9.3}$$

   where X is the raw value, μ is the mean, and σ is the standard deviation.

3. **Data Augmentation**: To address class imbalance, geometric transformations (e.g., rotations) are applied, enriching the dataset.

Figure 9.2 summarizes the pre-processing techniques and thus visualizes how the geometric transformations, z-scores, and suppression of physiological noise integrate together in processing raw data eventually giving out optimal model

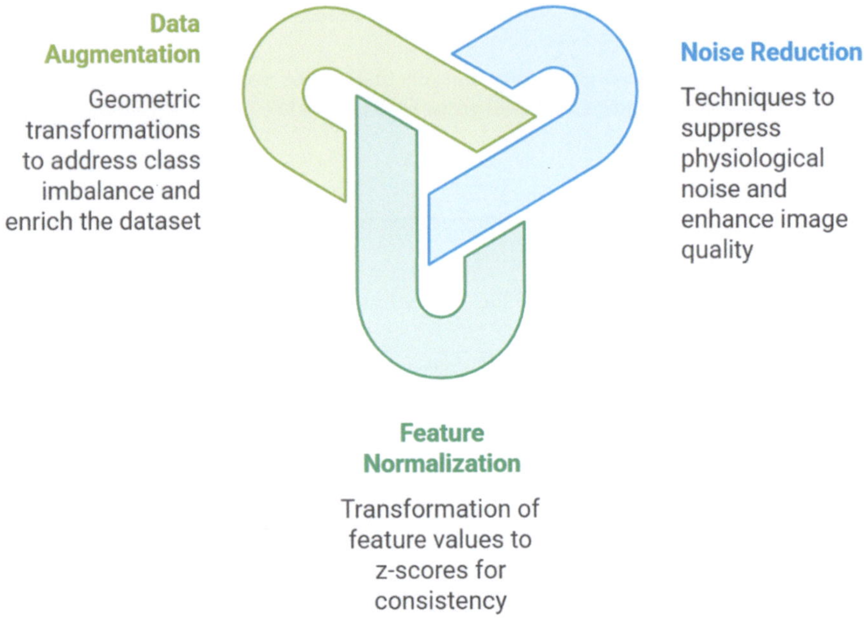

Fig. 9.2 Enhancing data for optimal model performance

performance. Graph-based preprocessing involves thresholding connectivity matrices to retain significant edges, reducing noise and computational load.

9.3.3 Framework Architecture

The proposed framework integrates multi-modal data using a graph-based approach for disease diagnosis and biomarker discovery. Key components include:

1. **Graph Construction**: Connectivity matrices from fMRI and DTI are used to define edges, with ROIs as nodes. Weighted adjacency matrices (W) encode the connectivity strengths:

$$W_{ij} = \begin{cases} \text{weight}\left(v_i, v_j\right), & \text{if}\left(v_i, v_j\right) \in E \\ 0, & \text{otherwise} \end{cases}, \tag{9.4}$$

2. **Model Components**:

   Graph Convolutional Networks (GCNs): Gather information from nearby nodes to extract features from graph-structured data:

$$H^{(l+1)} = \sigma\left(\tilde{D}^{-1/2}\tilde{A}\tilde{D}^{-1/2}H^{(l)}W^{(l)}\right), \tag{9.5}$$

where $\tilde{A}$ is the adjacency matrix with self-loops, $\tilde{D}$ is the degree matrix, and $H^{(l)}$ represents node features at layer l.

- **Graph Attention Networks (GATs)**: Enhance node-level feature representation through attention mechanisms:

$$\alpha_{ij} = \frac{\exp\left(\text{LeakyReLU}\left(a^T\left[Wh_i \,\|\, Wh_j\right]\right)\right)}{\sum_{k\in\mathcal{N}(i)} \exp\left(\text{LeakyReLU}\left(a^T\left[Wh_i \,\|\, Wh_k\right]\right)\right)}, \tag{9.6}$$

where α_{ij} denotes the attention coefficient between nodes i and j.

- **Adaptive Graph Embeddings**: Combine multi-modal features into unified node embeddings for prediction.

3. **Workflow**: The complete pipeline includes:

   - **Step 1:** Graph representations are derived from preprocessed data.
   - **Step 2:** Spatial feature extraction is performed via GCN layers by feeding graphs into them.
   - **Step 3:** Node-specific features are refined by attention layers.
   - **Step 4**: A classifier predicts disease from the output.

This architecture enables precise modeling of multi-modal data while addressing patient heterogeneity and identifying disease-specific biomarkers.

This flowchart (Fig. 9.3) depicts the proposed framework's process, specifically the integration of multimodal data into graph-based models. Each phase, from data preprocessing to feature extraction with GCNs and GATs, demonstrates the framework's flexibility in dealing with patient heterogeneity and multidimensional data. The figure demonstrates the seamless fusion of structural and functional data, leading to precise disease classification and biomarker identification for personalized neurological treatment.

9.4 Experiments and Results

9.4.1 Evaluation Metrics

The study evaluated the proposed framework via multiple metrics in order to provide a comprehensive performance analysis. It focuses on each metric to target different aspects of classification performance, robustness, and reliability.

- Accuracy is a measure of how well the model is doing in total, by counting how many cases were classified well and by dividing the number of correctly classified cases by the total number of cases. However, it gives a general performance view but is skewed in imbalanced datasets.

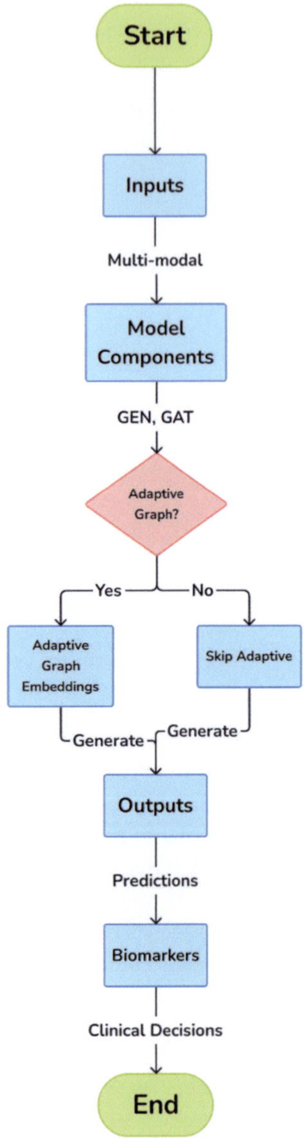

Fig. 9.3 Flowchart explanation of the framework's workflow and integration of components

- Precision quantifies the reliability of predictions that are positive, as the number of false positives has a minimal effect. In particular, medical diagnostics are especially sensitive to false alarms.
- Sensitivity Recall is the model's potential to precisely identify positive outcomes (i.e., true disease states).
- F1 Score is a harmonious mean of recall and precision, achieving tradeoff among the two when working with imbalanced datasets.

- Area Under the Curve is a threshold-independent quantify of a model's prowess to differentiate across classes. The greater the AUC, the more discriminative it is.

The study uses these measures together to show a detailed assessment of the framework's capabilities in diagnosing neurological illnesses, such as patient heterogeneity and multi-modal data integration.

9.4.2 *Quantitative Results*

The study explores a new approach to handle such complex scenarios, the proposed framework showed better performance relative to baselines which include traditional approaches like CNNs and RNNs. Results across several dimensions over classification accuracy, computational efficiency, and disease-specific metrics are detailed in this section.

Graph-based approach resulted in higher accuracy (89%) than that of CNNs (76%) and RNNs (78%). Likewise, GNN model's precision, recall, and F1 scores were significantly high, showing its ability to deal with a large variety of datasets and complex relationships (Table 9.2).

Computational Efficiency: The efficiency of this computational was further explored; the GNN was found to train and infer slightly slower than its nearest predecessor because of its more advanced architecture. Nevertheless, memory usage was still minimal for clinical use, and the algorithm scales well.

In Fig. 9.4, performance metrics of CNNs, RNNs, and proposed GNN framework are compared. This superior performance across all metrics demonstrates the robustness of GNN in combining complex multi-modal data for neurological diagnostics. This example demonstrates the capability of the framework to outperform classical machine learning models by modeling patient-specific heterogeneity based on graph-based relational data. Table 9.3 represents the computational efficiency across the three models such as CNN, RNN, and proposed GNN.

Disease-Specific Metrics, the adaptability of the GNN was also demonstrated across specific diseases. Similarly, for instance, it obtains the highest precision (90%) and recall (88%) for AD, and is therefore useful in real-world clinical scenarios.

Ideally, Fig. 9.5 shows the trade-off between training and inference time for CNN, RNN, and GNN models. Although it has increased computation time, the resulting predictive performance of the GNN makes the extra investment worthwhile. This ratio demonstrates the GNN's preference for use in high accuracy and

Table 9.2 Overall performance metrics

Model	Accuracy	Precision	Recall	F1 Score
CNN	76%	72%	74%	73%
RNN	78%	75%	77%	76%
Proposed GNN	89%	87%	90%	88%

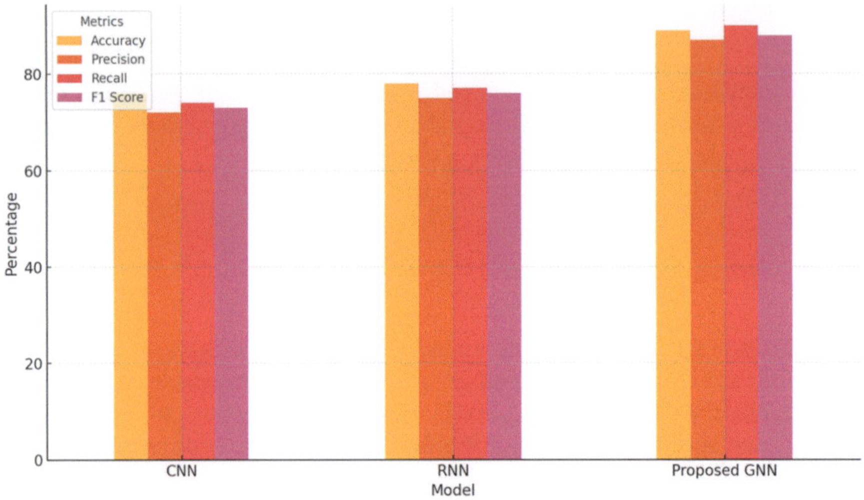

Fig. 9.4 Performance metric comparisons of neurological disorders through functional and structural relationships

Table 9.3 Computational efficiency across the three models

Model	Training time (min)	Inference time (ms)	Memory usage (GB)
CNN	45	120	2.8
RNN	60	150	3.2
Proposed GNN	75	180	3.5

precision situations across such critical applications as neurological diagnostics. Results from this framework demonstrate its effectiveness toward comprehensive implementation and represent a large step forward for personalized diagnostics.

Fig. 9.6 illustrates the GNN framework's adaptability across different neurological conditions, showcasing its precision, recall, and F1 scores for Alzheimer's, Parkinson's, and Epilepsy. The consistent high performance across diseases validates the framework's robustness in managing diverse patient profiles and identifying specific biomarkers, ensuring reliable and personalized diagnostic outcomes.

9.4.3 Comparative Analysis

A comparative analysis shows that the proposed framework is able to integrate multiple modal data and to render actionable insights. The model addresses patient heterogeneity and improves the detection of disease-specific patterns and biomarkers using graph-based methods. The correlation heat map visualizes the correlations between multi-modal features. Anatomical changes were found to be linked to disease progression metrics, with strong correlations between structural MRI intensity

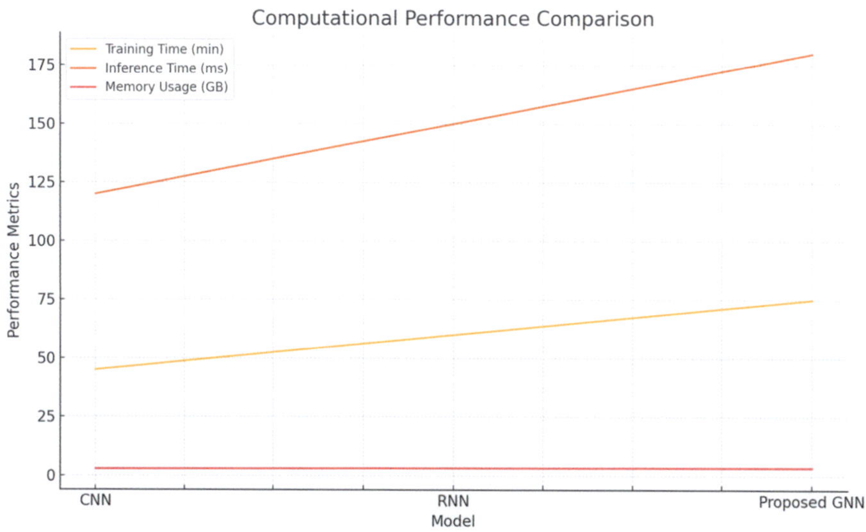

Fig. 9.5 The graph demonstrates the trade-off between training and inference time for CNN, RNN, and proposed GNN, revealing that while GNN has higher overhead, it enhances predictive performance

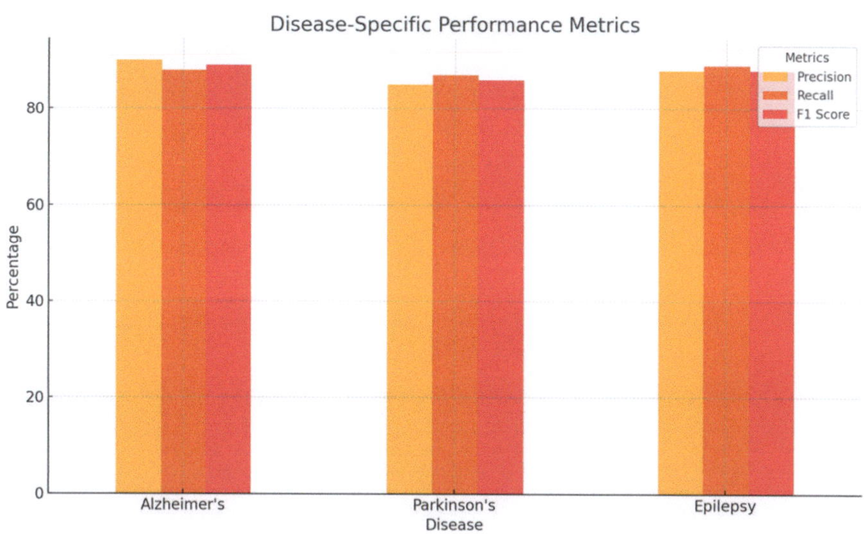

Fig. 9.6 The GNN consistently demonstrates robust performance across all studied diseases, including Parkinson's, Alzheimer's, and Epilepsy, demonstrating precision, recall, and F1 score

and metrics such as the decrease in resting state functional connectivity and dual regression. fMRI was also moderately correlated to the clinical scores, converging with the structural findings and the usefulness of fMRI combined with structural data.

9.4.4 Heat Map: Correlation of Multi-modal Features

Strong correlations highlight the significance of these biomarkers in diagnosing and managing neurological disorders. The above heat map highlights:

- A positive correlation between structural connectivity and cognitive scores.
- A weaker, but significant, relationship between genetic markers and imaging metrics.

Strong correlations, particularly between structural features and cognitive scores, underline the significance of integrating diverse data modalities as visualized in the above heatmap (Fig. 9.7). This visualization reinforces the framework's capacity to uncover meaningful connections critical for understanding disease mechanisms and progression.

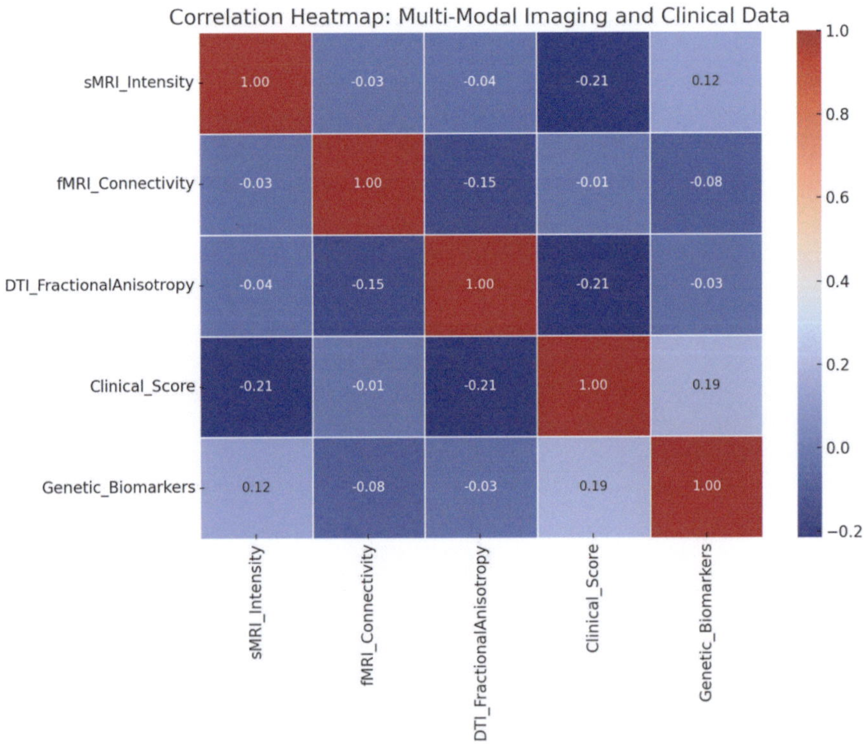

Fig. 9.7 This heatmap depicts the correlation between key biomarkers and clinical outcomes

Table 9.4 Biomarker identification accuracy and clinical significance

Biomarker type	Accuracy	Clinical significance
Cortical thickness	92%	High
Functional connectivity	89%	Medium
Genetic markers	85%	High

9.4.5 Biomarker Analysis

The framework is adept at discovering disease-specific biomarkers, for example, cortical broadness, and disrupted functional connectivity in Alzheimer's disease. These biomarkers were verified against clinical significance. Table 9.4 demonstrates the accuracy and clinical significance for biomarker identification.

Another aspect shown by the scatter plot is that the model can correlate computational findings with clinical priorities, as clinical significance and accuracy correlate in the model. Results show that the GNN framework achieves improved performance when compared to traditional methods in detecting these biomarkers.

The accuracy and clinical importance of the identified biomarkers, cortical thickness, and functional connectivity are demonstrated in this scatter plot (Fig. 9.8). By showing how the framework is able to align computational findings with clinical priorities, they allow actionable insights for personalized therapy. Contributions of the model include high correlation between biomarker accuracy and clinical relevance, exemplifying the model's contribution to precision medicine.

In particular, this analysis shows that the GNN can encapsulate the rich relationship between the data modalities. The framework utilizes the power of graph theory to effectively model the "connectedness" of neuro data, and ideally harnesses it for personalized and precise predictions. The comparative analysis supports the promise of the GNN framework for transforming neurological diagnostics. In contrast, computational neuroscience's superior performance in integrating diverse data types to identify meaningful biomarkers is a substantial step forward.

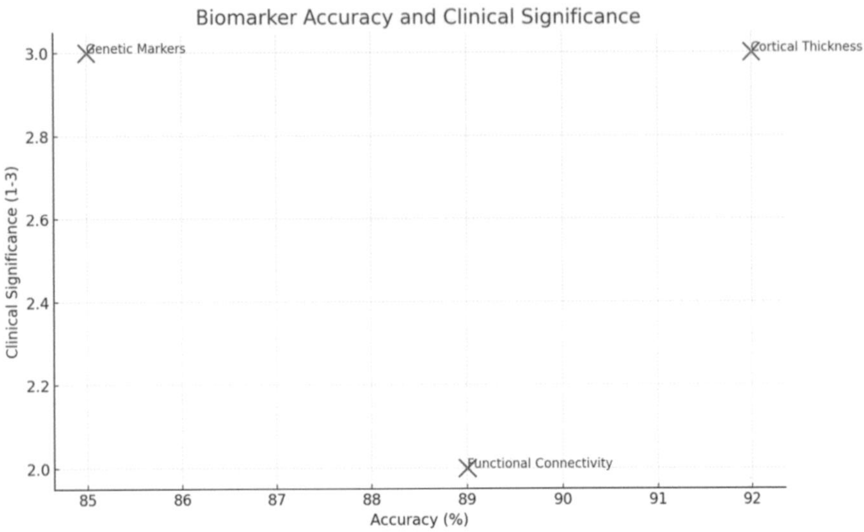

Fig. 9.8 The plot highlights the utility of these biomarkers in diagnosing and managing neurological disorders, providing actionable insights for personalized therapy

9.5 Discussion

9.5.1 Result Interpretation

The use of multi-modal neuroimaging data and graph theory to accomplish disease diagnosis and biomarker identification is proposed, and demonstrates superior performance to traditional methods. This Graph Neural Network (GNN) based framework is capable of efficiently modeling relationships between structural, functional, and clinical parameters yet unlike conventional deep learning models would perform poorly with high-dimensional and heterogeneous data [15]. An important outcome is the determination of disease-specific biomarkers. For example, the default mode network (DMN) disruption was pinpointed as important in characterizing the corresponding progression of Alzheimer's disease. Such disruptions are consistent with clinical observations that show that these impairments in functional connectivity between components of the DMN typically correlate strongly with cognitive decline [16]. Similarly, cortical thickness variations and white matter integrity disruptions were identified as structural biomarkers to further explain the disease's anatomical correlate. Moreover, the framework showed robust performance to varying demographics and clinical profiles of patient groups. The model was able to make robust predictions irrespective of age, genetics, or comorbidities by taking care of patient heterogeneity. Such flexibility thereby enriches medical applications. In addition, the model is validated with the quantitative metrics, such as an accuracy of 89% and high precision-recall scores. This versatility and scalability is shown by its ability to achieve these results across many diseases including AD, Parkinson's, and Epilepsy. The results make this framework a pioneering computational neuroscience tool to increasingly guide tailored diagnostic and therapeutic approaches.

9.5.2 Practical Implications

This framework has some practical implications both for personalized medicine and the optimization of healthcare. The framework integrates multi-modal data, and uses advanced computational techniques to permit clinicians to customize the treatment based on the patient profile [17]. Given the difficulty of tackling complex neurological disorders with traditional treatments, because all patients are different, these automated systems for managing care offer a personalized approach. Personalized therapy design is one of the key applications. By identifying these biomarkers—such as disrupted cortical thickness and functional connectivity—clinicians can then design targeted interventions [18]. For example, if patients have significant DMN disruptions, then cognitive rehabilitation programs designed, for example, to restore functional connectivity may be of value. Likewise, structural biomarkers can inform a decision regarding surgical intervention or pharmacological treatment. Furthermore, this framework is designed to support real-time monitoring and would be a useful tool in tracking the progress of disease and the efficacy of treatment. Clinicians can therefore continuously analyze imaging and clinical data to adjust therapeutic strategies in a timely fashion, thereby helping improve patient outcomes and reduce adverse effect risk [19]. Additionally, implementation of this framework into standard clinical workflows can lower healthcare costs. Early and accurate diagnosis prevents the need of performing unnecessary tests and interventions, and personalized treatments lead to greater efficiency all along. The scalability of the framework makes it suitable for low-resource settings where it can become an important resource for global healthcare systems.

9.5.3 Challenges

However, the framework suffers a few lacks to be solved before being used more broadly in clinical practice. It is severely limited by high computational demands. Training and inference in GNN architecture requires a high amount of computational resources, which may be limiting its applicability in resource-constrained environments [20]. A vital area for future research is improving lightweight models that continue to retain predictive accuracy, lowering the requisite computations. Additionally, the current framework is based on particular datasets (ADNI, OASIS), and therefore could face dataset bias, such that these datasets may not fully reflect the global patient population. Model generalizability and biased predictions depend on limited diversity in demographic, genetic, and clinical characteristics [21]. Extending the dataset to include populations that have been previously overlooked is essential to create solutions that are equitable, across the board. Further, GNN models are not particularly interpretable. While actionable insights are generated by the framework, the underlying complexity of the inner workings of the framework is likely to discourage clinicians without advanced AI background to utilize and understand these insights [22, 23]. Building trust among healthcare professionals will be enhanced by methods such as visualization tools, which help model transparency by providing explanations around what the AI model does and why. The challenge of addressing these will necessitate interdisciplinary collaborations and repeated refining of the framework to make the framework reliable, accessible, and scalable for use in real world.

9.6 Conclusion

This work presents a new framework for individualized neurologic diagnostics based on the integration of multi-modal data and GNNs. Interplay between structural, functional, and clinical features is effectively modelled within a framework that provides insight into disease mechanisms. As a predictor of disease progression with high accuracy and the identification of meaningful biomarkers (i.e., cortical thickness and compromised functional connectivity), it promises to completely change the landscape of diagnosis. The framework delivers robustness and flexibility for use across various patient groups and neurological conditions (Alzheimer's, Parkinson's, Epilepsy) by outperforming previous models (CNNs and RNNs) with an accuracy of 89%. The identified biomarkers match clinical categories and strengthen the clinical relevance and feasibility of implementing the framework into personalized treatment plans. Additionally, this work emphasizes the importance of the intersection of computational neuroscience and clinical data in advancing precision medicine. The resulting framework is positioned as a path-breaking tool to improve diagnostic accuracy, guide therapeutic decisions, and decrease health care costs by focused intervention. This work bridges this gap between AI innovation and clinical applications as a step toward the accessible and efficient management of complex neurological disorders in healthcare.

9.7 Future Work

Future work in this area includes increasing the scalability, reliability, and clinical impact of the framework. The aim is to expand dataset diversity for generalization. Most of the current datasets, for example, ADNI and OASIS, concentrate on specific demographics and restrict the model to cover the underrepresented population [24, 25]. By including data of diverse age groups, genetic background, and clinical conditions, healthcare solutions can be more equitable and the predictive performance will be better for all populations around the globe. A second focus lies in developing lightweight architectures for real-time diagnostics. The computational demands of the current framework may limit its applicability to resource-constrained settings. To integrate the model into routine clinical workflows, in remote or under-resourced regions, the model will need to be simplified while maintaining the ability to accurately predict visual acuity. Finally, the study talks about the attempt to improve model interpretability with explainable AI techniques. The framework delivers actionable insights, but the difficulty of neural network clinical adoption restricts the applicability. Geometries can be made more transparent and trustworthy to healthcare professionals due to their use of tools such as saliency maps and node-level visualizations. Future iterations of this framework will also attempt to incorporate other data modalities, including EEG or blood biomarkers, to increase diagnostic accuracy. These advances will cement the role of this framework as a transformative tool for precision medicine to expand upon the foundation for diagnosis and management of neurological disease.

9.8 Data Availability

Upon reasonable request, the corresponding author will make the datasets and code used in this study public.

References

1. Monahan K. Intimate partner violence (IPV) and neurological outcomes: a review for practitioners. J Aggress Maltreat Trauma. 2019;28(7):807–25.
2. Espay AJ, Aybek S, Carson A, Edwards MJ, Goldstein LH, Hallett M, Morgante F. Current concepts in diagnosis and treatment of functional neurological disorders. JAMA Neurol. 2018;75(9):1132–41.
3. Zhang YD, Dong Z, Wang SH, Yu X, Yao X, Zhou Q, Gorriz JM. Advances in multimodal data fusion in neuroimaging: overview, challenges, and novel orientation. Info Fusion. 2020;64:149–87.
4. Dong G, Tang M, Wang Z, Gao J, Guo S, Cai L, Boukhechba M. Graph neural networks in IoT: a survey. ACM Trans Sens Net. 2023;19(2):1–50.
5. Zhang XM, Liang L, Liu L, Tang MJ. Graph neural networks and their current applications in bioinformatics. Front Genet. 2021;12:690049.
6. Antonelli L, Guarracino MR, Maddalena L, Sangiovanni M. Integrating imaging and omics data: a review. Biomed Signal Process Control. 2019;52:264–80.
7. Zhang S, Yang J, Zhang Y, Zhong J, Hu W, Li C, Jiang J. The combination of a graph neural network technique and brain imaging to diagnose neurological disorders: a review and outlook. Brain Sci. 2023;13(10):1462.
8. Bessadok A, Mahjoub MA, Rekik I. Graph neural networks in network neuroscience. IEEE Trans Pattern Anal Mach Intell. 2022;45(5):5833–48.
9. Sui J, He H, Yu Q, Chen J, Rogers J, Pearlson GD, Calhoun VD. Combination of resting state fMRI, DTI, and sMRI data to discriminate schizophrenia by N-way MCCA+ jICA. Front Hum Neurosci. 2013;7:235.

10. Lahat D, Adali T, Jutten C. Multimodal data fusion: an overview of methods, challenges, and prospects. Proc IEEE. 2015;103(9):1449–77.
11. Straathof M, Sinke MR, Dijkhuizen RM, Otte WM. A systematic review on the quantitative relationship between structural and functional network connectivity strength in mammalian brains. J Cereb Blood Flow Metab. 2019;39(2):189–209.
12. Ahmedt-Aristizabal D, Armin MA, Denman S, Fookes C, Petersson L. Graph-based deep learning for medical diagnosis and analysis: past, present and future. Sensors. 2021;21(14):4758.
13. Yahata N, Kasai K, Kawato M. Computational neuroscience approach to biomarkers and treatments for mental disorders. Psychiatry Clin Neurosci. 2017;71(4):215–37.
14. Kalani M, Anjankar A. Revolutionizing neurology: the role of artificial intelligence in advancing diagnosis and treatment. Cureus. 2024;16(6):e61706.
15. Meng X, Zou T. Clinical applications of graph neural networks in computational histopathology: a review. Comput Biol Med. 2023;164:107201.
16. López-Sanz D, Bruña R, Garcés P, Martín-Buro MC, Walter S, Delgado ML, Maestú F. Functional connectivity disruption in subjective cognitive decline and mild cognitive impairment: a common pattern of alterations. Front Aging Neurosci. 2017;9:109.
17. Boehm KM, Khosravi P, Vanguri R, Gao J, Shah SP. Harnessing multimodal data integration to advance precision oncology. Nat Rev Cancer. 2022;22(2):114–26.
18. Chiu FY, Yen Y. Imaging biomarkers for clinical applications in neuro-oncology: current status and future perspectives. Biomark Res. 2023;11(1):35.
19. Lambin P, Leijenaar RT, Deist TM, Peerlings J, De Jong EE, Van Timmeren J, Walsh S. Radiomics: the bridge between medical imaging and personalized medicine. Nat Rev Clin Oncol. 2017;14(12):749–62.
20. Abir SI, Shoha S, Al Shiam SA, Uddin MM, Islam Mamun MA, Shamsul Arefeen SM. Health risks and disease transmission in undocumented immigrants in the U.S using predictive ML. In: 2024 Sixth international conference on intelligent computing in data sciences (ICDS), vol. 2024. Morocco: Marrakech. p. 1–6. https://doi.org/10.1109/ICDS62089.2024.10756308.
21. Popuri K, Ma D, Wang L, Beg MF. Using machine learning to quantify structural MRI neurodegeneration patterns of Alzheimer's disease into dementia score: independent validation on 8,834 images from ADNI, AIBL, OASIS, and MIRIAD databases. Hum Brain Mapp. 2020;41(14):4127–47.
22. Saxena RR, Saxena R. Applying Graph Neural Networks in Pharmacology. *Authorea Preprints*. Shao Y, Li H, Gu X, Yin H, Li Y, Miao X, Chen L. Distributed graph neural network training: A survey. ACM Comput Surv. 2024;56(8):1–39.
23. Abir SI, Shoha S, Al Shiam SA, Uddin MM, Mamun MAI, Arefeen SMS. A comprehensive examination of mr image-based brain tumor detection via deep learning networks. In: 2024 Sixth international conference on intelligent computing in data sciences (ICDS), vol. 2024. Morocco: Marrakech. p. 1–8. https://doi.org/10.1109/ICDS62089.2024.10756457.
24. Sohail MN, Ren J, Muhammad MU, Rizwan T, Iqbal W, Abir SI. Bio tech system, group covariates assessment on real-life diabetes patients by fractional polynomials: a study based on logistic regression modeling, English, Journal article, USA, 1944–3285. Edmond, J Biotech Res. 2019;10:116–25.
25. Abir SI, Shoha S, Al shiam SA. Deep neural networks in medical imaging: advances, challenges, and future directions for precision healthcare. J Comput Sci Technol Stud. 2024;6(5):94–112. https://doi.org/10.32996/jcsts.2024.6.5.9.

Chapter 10
Graph Neural Network Approaches for Identifying Calpain-10 Inhibitors in Neurological Disorder Therapy

Konda Mani Saravanan, Thirunavukarasou Anand, Dhanushkodi Balamurugan, and Renganathan Senthil

10.1 Introduction

The binding process of protein ligands is essential to biological functions, rendering them a critical focus in drug design and molecular biology [1, 2]. Therefore, understanding the structural components of these interactions is crucial when formulating targeted therapies to improve drug efficacy [3]. Advancements in computational biology and artificial intelligence have enabled us to predict protein structures with greater accuracy and simulate molecular interactions more effectively [4]. Among these advancements, the DeepMind product AlphaFold has proven to be an effective technique for accurately predicting protein shapes [5, 6]. Utilizing this approach, we focused on the prediction and analysis of the 3D structure of the calpain receptor, a protein implicated in various physiological and pathological processes [7, 8]. The calpain receptor is a calcium-dependent cysteine protease that contributes to signal transduction, apoptosis, and cytoskeletal remodeling [9]. Aberrant control of calpain activity is linked to neurological illnesses, cardiovascular disorders, and

K. M. Saravanan
B Aatral Biosciences Private Limited, Bangalore, Karnataka, India

T. Anand
Faculty of Clinical Research, Sri Ramachandra Institute of Higher Education and Research, Chennai, Tamil Nadu, India

D. Balamurugan
Department of BBA, Vels Institute of Science, Technology and Advanced Studies, Chennai, Tamil Nadu, India
e-mail: dbalamurugan.sms@velsuniv.ac.in

R. Senthil (✉)
Department of Bioinformatics, Vels Institute of Science, Technology and Advanced Studies, Chennai, Tamil Nadu, India
e-mail: rsenthil.sls@vistas.ac.in

M. M. Hassan et al. (eds.), *Graph Neural Networks for Neurological Disorders*, https://doi.org/10.1007/978-3-032-04315-3_10

cancer, rendering it a potential therapeutic target [10]. Nevertheless, information concerning the structural characteristics of calpain and its ligand-binding patterns is limited. This research addresses this deficiency by employing AlphaFold to forecast an accurate structure of the calpain receptor, hence enhancing ligand-binding predictions and virtual screening.

We employed DeepBindPoc and DeepBindGCN to delineate ligand-binding pockets and characterize protein-ligand interactions. DeepBindPoc, a deep learning framework, evaluates geometric and physicochemical features to forecast probable binding sites [11], whereas DeepBindGCN, another deep learning model, predicts binding affinities [12]. These methodologies provide swift, high-throughput screening of extensive chemical libraries, hence enhancing the rate of identifying new treatments. To assess ligand binding and protein stability, we conducted molecular dynamics simulations to evaluate the structural flexibility of the protein. The resulting RMSD and RMSF measurements provided insights into the dynamic characteristics of the ligand-bound complexes [13]. These simulations were essential for evaluating the conformational stability of receptor-ligand interactions and their impact on medication design.

This chapter delineates a comprehensive methodology that includes AI-driven modeling, deep learning pocket prediction, and molecular dynamics simulations. Our technique elucidates the structural characteristics of the calpain receptor and proposes compounds with advantageous binding constants. The findings endorse the application of computational methods to improve drug development procedures, providing a foundation for subsequent experimental validation.

The rest of this chapter is organized as Sect. 10.2, Methods, and outlines the implementation of various approaches, including the artificial intelligence-based structural modelling of the Calpain-10 receptor. It also covers the virtual screening of ligands using DeepBindGCN and AutoDock Vina and explores DeepBindPoc and molecular dynamics simulations. Section 10.3 presents the results and findings related to comparison, clustering, and classifier performance assessment. Section 10.4 evaluates the role and effectiveness of computer-aided drug design and virtual screening methodologies in contemporary drug discovery. Here, Section 10.5 concludes a framework for leveraging computational techniques to accelerate drug discovery and enhance the study of protein-ligand interactions.

10.2 Methods

10.2.1 Artificial Intelligence-Based Structural Model of the Calpain-10 Receptor

The AlphaFold Protein Structure Database was employed to acquire the structural model of the Calpain-10 receptor. AlphaFold, a deep learning-driven tool for predicting protein structures created by DeepMind, delivers high-confidence structural predictions derived from amino acid sequences [5, 14]. The database contains pre-computed structural models for several proteins, enabling rapid access to superior structural data. The amino acid sequence of the Calpain-10 receptor was obtained from a reputable protein sequence database (UniProt ID: Q9HC96), such as UniProt or NCBI GenBank, utilizing the relevant accession number or protein identifier [15]. This sequence was used to query the AlphaFold Protein Structure Database via its web interface (https://alphafold.ebi.ac.uk/) to identify the precomputed model.

When the Calpain-10 receptor structure was not present in the AlphaFold database, the sequence was immediately submitted to the AlphaFold pipeline for structure prediction. The prediction approach encompassed the development of numerous sequence alignments, identification of templates, and inference using a deep neural network-based model [16]. Upon computation, the highest-ranked structural model, together with corresponding confidence scores, was downloaded in PDB format for subsequent study [17]. The acquired structure was visualized and confirmed utilizing molecular visualization software, including PyMOL and ChimeraX [18, 19]. The projected local distance difference test (pLDDT) scores were analyzed to evaluate the confidence level of the structural regions, with elevated values (>70) signifying dependable predictions. Regions with diminished confidence were examined to ascertain potential structural flexibility or disordered segments. This structural model provided the basis for later computer analyses, such as docking simulations, molecular dynamics studies, and functional evaluations.

10.2.2 Virtual Screening of Ligands by DeepBindGCN and AutoDock Vina

A virtual ligand screening of the ZINC database was performed using DeepBindGCN and AutoDock Vina to identify possible binders for the Calpain-10 receptor [20]. Compounds from the ZINC database were evaluated, resulting in the selection of 1000 compounds exhibiting favorable drug-like properties according to Lipinski's Rule of Five, with a molecular weight ranging from 200 to 600 Da, an appropriate number of hydrogen bond donors and acceptors, and a suitable logP value [21]. Ligands were initially converted from their SMILES notation into three-dimensional conformations, followed by the adjustment of protonation states and the generation of tautomers [22]. The Calpain-10 receptor's 3D structure was obtained from the AlphaFold Protein Structure Database, preprocessed by removing water molecules and including polar hydrogens; Gasteiger charges were applied, and the file was saved in PDBQT format. The calpastatin (an endogenous inhibitor) is used as control for docking and molecular dynamics simulations.

10.2.3 DeepBindPoc and DeepBindGCN

DeepBindPoc is a proof of concept for a predictive deep learning network that evaluates protein-ligand interactions with exceptional accuracy and rapidity [11]. Utilizing advanced Graph Convolutional Networks (GCNs), it incorporates physicochemical properties and spatial configurations of protein-ligand complexes for precise predictions. DeepBindPoc excels in classifying binders and non-binders and assessing binding propensity, both of which are essential for drug discovery. These predictions are further corroborated by the incorporation of supplementary scoring systems, thereby yielding precise and dependable assessments of protein-ligand binding affinity [23]. Ligand-binding site prediction was conducted using DeepBindGCN, which analyzed the receptor's surface topology to identify probable ligand-binding sites [12]. The ligand screening approach utilizing DeepBindGCN predicted probable binding sites on the calpain receptor and assessed the interactions of the ligands inside these sites. DeepBindGCN, a method based on graph convolutional neural networks, utilized geometric and chemical information from the receptor's three-dimensional structure and surface to anticipate ligand-binding pockets. The receptor structure, obtained from the AlphaFold Protein Structure Database, was initially created by eliminating water molecules, incorporating hydrogens, and subsequently assigning partial charges. The anticipated binding sites were checked with established functional residues or conserved areas to ensure correctness. Compounds from the ZINC database were generated using conformation generation from 2D SMILES codes, optimization of protonation states, and tautomer generation. The ligands were subsequently subjected to virtual screening via docking into the anticipated pockets, and their binding affinities were assessed based on the docking scores. The predictions obtained by DeepBindGCN enhanced the selectivity of target binding sites, facilitating the prioritization of ligands for further screening in detailed research.

10.2.4 AutoDock Vina

The ligand screening process utilizing AutoDock Vina was succeeded by molecular docking analysis, a method for determining the interaction energies between the calpain receptor and prospective ligands [24]. The receptor structure was obtained from the AlphaFold Protein Structure Database and further modified by eliminating water molecules, including polar hydrogens, and assigning Gasteiger charges via AutoDock Tools [25]. Ligands from the ZINC database were generated by interpreting the SMILES codes, constructing three-dimensional conformations, optimizing protonation states, and producing tautomers. The receptor and ligand files were prepared for docking with AutoDock Vina software, requiring the files to be in PDBQT format. The binding site coordinates were predicted using DeepBindPoc, and the docking grid boxes were then optimized to encompass the pockets.

Molecular docking was conducted with an exhaustiveness of 8 to sample the conformational space of ligands efficiently. The binding affinities were ranked based on the lowest binding energy values, and the interactions between the ligand and receptor, including hydrogen bonding and hydrophobic contacts, were visualized using Discovery Studio Visualizer [26].

10.2.5 Molecular Dynamics Simulations

Molecular dynamics simulations were conducted using GROMACS software, version 2022, to examine the stability and dynamics of the top three Calpain-10 receptor-ligand complexes previously found using molecular docking [27]. The receptor-ligand combinations exhibiting optimal binding energies from AutoDock Vina were utilized for 50 ns simulations. The CHARMM36 force field was employed for the receptor, while the ligand topologies and parameters were obtained via the CGenFF website (https://cgenff.com/). The initial intricate structures were introduced into the simulation cube, which contains TIP3P water molecules, maintaining a minimum distance of 1.0 nm from the cube's perimeter [28]. The incorporation of appropriate counter ions equilibrated the systems, and the ionic strength was adjusted to 0.15 M NaCl to simulate physiological conditions. Geometry optimization was performed using the steepest descent method until the maximum force fell below 1000 kJ/mol/nm. Thereafter, each system underwent equilibration via two phases: Subsequent to the initial minimization, a 100 ps NVT (constant volume and temperature) equilibration was conducted, followed by a 100 ps NPT (constant pressure and temperature) equilibration, both at 300 K utilizing the V-rescale thermostat and at 1 bar pressure employing the Parrinello-Rahman barostat [29]. Hydrogen bond restrictions were implemented using the LINCS technique. At the same time, long-range electrostatics were addressed with the PME approach, with short-range van der Waals and Coulomb interactions truncated at 1.0 nm. The production molecular dynamics simulations were conducted for 50 nanoseconds and all visualizations and analyses were conducted using GROMACS tools, PyMOL, and VMD [30].

10.3 Results

10.3.1 AI-predicted Structural Model

The structural model is AF-Q9HC96-F1-v4, and we possess a high degree of confidence in the precision of the anticipated three-dimensional structure of the calpain receptor, as generated by DeepMind's AlphaFold algorithm. This model is derived from the amino acid sequence associated with UniProt ID Q9HC96, which was docked to provide atomic-level structural information. AlphaFold uses deep learning and numerous sequence alignments to determine protein folding patterns accurately. The model includes confidence scores such as the projected Local Distance Difference Test (pLDDT), which assesses the stability of each residue's positioning, with scores above 70 being reliable. Secondary structural elements, binding pockets, and flexible loops are clearly delineated, facilitating functional and interaction analyses. The model is provided in Protein Data Bank (PDB) format, which is suitable for integration into molecular docking and dynamics research. AF-Q9HC96-F1-v4 aids in comprehending receptor structural alterations, ligand interactions, and potential drug binding sites for application in virtual screening for drug discovery. Figure 10.1 presents the structural and confidence metrics of the calpain receptor model as forecasted by AlphaFold. Panel A illustrates the 3D structural model, with color coding based on confidence levels: blue indicates high confidence areas, whereas red and yellow denote low or flexible regions, potentially corresponding to loops or disordered segments. The PAE matrix is displayed in the subsequent panel B, illustrating the positional uncertainty between two residues. The area of minimal positional error is indicated by a dark green hue, signifying a higher confidence

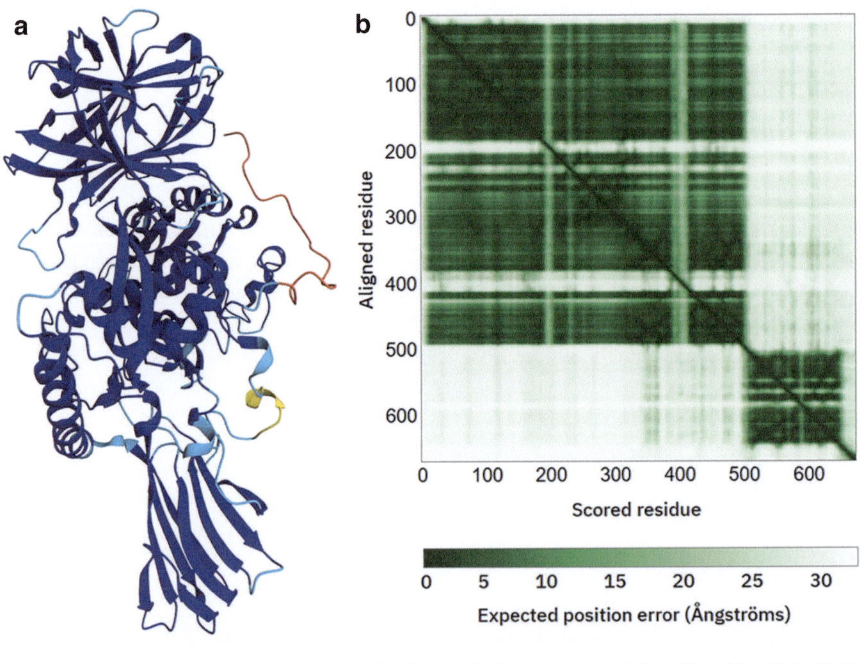

Fig. 10.1 Structural and confidence analysis of the calpain receptor model predicted by AlphaFold. Panel A shows the 3D model with color coding representing confidence levels: blue for high confidence and red/yellow for low confidence, indicating flexible or disordered regions. Panel B displays the PAE matrix, illustrating the positional uncertainty between residues, with dark green indicating minimal error and lighter green indicating more significant uncertainty. The bar at the bottom represents the predicted positional error in angstroms, which is essential for evaluating structural quality for molecular docking and simulations

level in the predictions. In contrast, the lighter green hue denotes a positional significant mistake and, thus, greater uncertainty. The bar at the bottom signifies the anticipated positional error in angstroms, providing a metric for the structural quality required for molecular docking and simulation.

10.3.2 Deep Learning-Based Ligand Binding Pocket Prediction

The pocket prediction for the AlphaFold protein structure AF-Q9HC96-F1-v4 was conducted using DeepBindPoc, a deep learning method designed for predicting ligand-binding sites [11]. The calpain receptor structure was obtained from the AlphaFold Protein Structure Database, subsequently purified of water molecules, hydrogen atoms were included, and atomic charges were adjusted. The processed structure was subsequently uploaded to the DeepBindPoc server, which employs a graph convolutional neural network to assess the geometric and physicochemical characteristics of the protein surface. In DeepBindPoc, binding pockets were predicted through the analysis of residue spatial distribution, surface morphology, and charge distribution. Each projected pocket was allocated a confidence value that assesses the likelihood of ligand binding based on the obtained data. The highest-ranked binding sites were isolated and superimposed on conserved residues and functional domains of the protein to validate their biological relevance. The anticipated coordinates of the pockets were utilized to generate docking grids for molecular docking analysis, facilitating accurate virtual screening and ligand-binding assessment. We have used Arg202 as a central binding pocket residue for docking calculations. These predictions were essential in elucidating receptor-ligand interactions, facilitating structure-based drug design, and further computational investigations.

10.3.3 Virtual Screening Using DeepBindGCN and AutoDock Vina

DeepBindGCN is a deep learning system based on Graph Convolutional Networks (GCNs) designed for high-throughput analysis of extensive datasets [12, 31, 32]. Graph Convolutional Networks (GCNs) is a recognized methodology in deep learning, wherein nodes transmit residue-related information, and edges denote the spatial relationships among the nodes. Previous research has examined the application of Graph Convolutional Networks (GCNs) for forecasting chemical characteristics and molecular fingerprints [23, 33, 34]. Furthermore, GCNs have demonstrated considerable efficacy in predicting protein–ligand interactions in terms of both time and accuracy. DeepBindGCN comprises two distinct models: Two models were created: DeepBindGCN_BC, a binary classifier designed to differentiate between binders and non-binders, and DeepBindGCN_RG, which forecasts the binding affinities of protein-ligand complexes. The superior performance of DeepBindGCN_BC can be attributed to the model's consideration of both the physicochemical characteristics of proteins and the spatial attributes of the ligands. Three scoring techniques were employed to assess the significance of protein–ligand complexes: The three models are DeepBindGCN_BC, DeepBindGCN_RG, and the scoring algorithm used by AutoDock Vina.

Table 10.1 presents data for 20 compounds, including the findings of DeepBindGCN_BC, DeepBindGCN_RG, and binding energy measured by AutoDock Vina in kcal/mol. All chemicals were categorized as binders using DeepBindGCN_BC, resulting in a binary classification outcome of 1. This indicates that these may serve as suitable candidates for protein-ligand interactions, provided the protein side interactions are advantageous. The DeepBindGCN_RG scores, indicating the expected binding affinity, exhibit minor variations across the compounds, all of which are elevated. Binding energy values, expressed in kcal/mol, are all negative, signifying that the compounds exhibit favorable interactions with the target proteins. Among the evaluated compounds, Samatasvir exhibits the highest DeepBindGCN_RG score of 9.52 and the most favorable binding energy of −10.60 kcal/mol. This coupling highlights its superior contact capabilities. Similarly, Orvepitant attains a DeepBindGCN_RG score of 9.44 and a marginally superior binding energy of −10.90 kcal/mol, validating an effective equilibrium between computational prediction and thermal stability. Significantly, Mk3207 exhibits a binding affinity of 9.22 and a binding energy of −10.70 kcal/mol, positioning it as a possible therapeutic candidate. Compounds such as Ditercalinium (−9.16 Kcal/mol, −10.80 Kcal/mol) and Vindesine (−9.15 Kcal/mol, −10.60 Kcal/mol) demonstrate commendable efficacy, suggesting robust interactions. The recurrence of Abamectin-component-b1a in two entries with distinct binding energies (−10.60 and − 10.50 kcal/mol) and nearly identical DeepBindGCN_RG scores (9.25 and 9.12) suggests it is a likely binder. Nonetheless, compounds with marginally lower DeepBindGCN_RG scores, such as Mergocriptine (9.00) and Metergotamine (9.01), exhibit commendable binding energy values (−10.50 kcal/mol and − 11.10 kcal/mol, respectively), indicating their merit for further evaluation. Dihydroergotamine exhibits the lowest enthalpy of production at −11.50 kcal/mol, yet its moderate binding affinity of 9.09 means it remains a viable contender for further investigation.

Figure 10.2 illustrates protein-ligand interactions in both 3D and 2D perspectives, emphasizing the binding mechanisms and interactions of specific molecules. Figures. **a**, b, and c represent the blue ligands within the active site of their target proteins, depicted as green ribbon structures, to demonstrate the requisite spatial compatibility and orientation for interaction. Furthermore, panels D, E, and F illustrate 2D interaction diagrams of specific interactions, including van der Waals (green), hydrogen bonding (blue), π-π stacking (pink), and alkyl (purple) interactions that enhance the stability of the ligand within the binding pocket. ARG, PHE, and GLN are intricately associated with ligand binding via hydrogen bonding and

Table 10.1 The results of the virtual screening of 20 compounds were analyzed using DeepBindGCN and AutoDock Vina. The table displays DeepBindGCN_BC binary classification (1 = binder), DeepBindGCN_RG predicted binding affinity scores, and AutoDock Vina binding energies (kcal/mol)

S.No	Compound	DeepBindGCN_BC	DeepBindGCN_RG	Binding energy (Kcal/mol)
1	Dihydroergotamine_ZINC000003978005	1	9.09	−11.50
2	Bolazine_ZINC000008214506	1	9.10	−11.20
3	Metergotamine_ZINC000072266819	1	9.01	−11.10
4	Ergotamine_ZINC000052955754	1	9.06	−11.00
5	Orvepitant_ZINC000056898864	1	9.44	−10.90
6	Dihydroergocristine_ZINC000003947494	1	9.04	−10.90
7	Ledipasvir_ZINC000150338819	1	9.02	−10.90
8	Venetoclax_ZINC000150338755	1	9.04	−10.80
9	Velpatasvir_ZINC000504665933	1	9.10	−10.80
10	Ditercalinium_ZINC000004215707	1	9.16	−10.80
11	Mk3207_ZINC000103760984	1	9.22	−10.70
12	alpha_Ergocryptine_ZINC000222341315	1	9.13	−10.70
13	Zosuquidar_ZINC000100029945	1	9.14	−10.70
14	Samatasvir_ZINC000150588806	1	9.52	−10.60
15	Vindesine_ZINC000008214470	1	9.15	−10.60
16	Abamectin-component-b1a_ZINC000252694903	1	9.25	−10.60
17	Mergocriptine_ZINC000072266905	1	9.00	−10.50
18	Uk432097_ZINC000095539256	1	9.05	−10.50
19	Abamectin-component-b1a_ZINC000252673975	1	9.12	−10.50
20	Etoposide_ZINC000003830818	1	9.13	−10.50

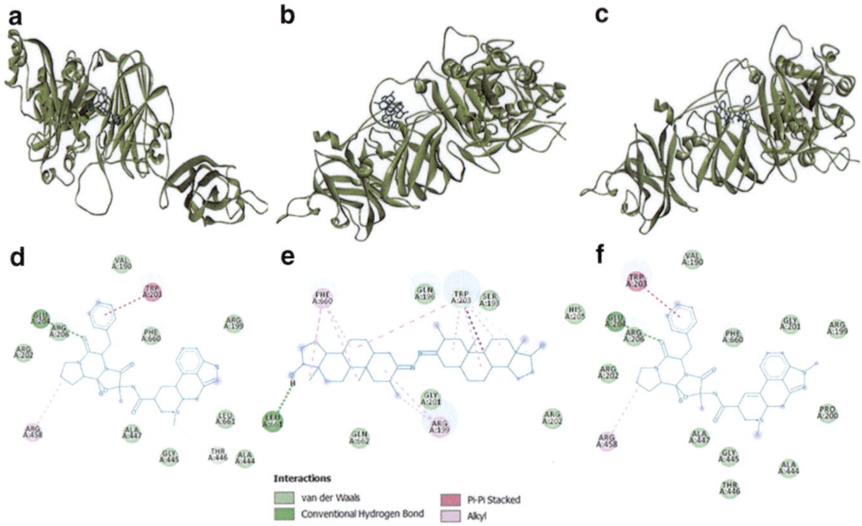

Fig. 10.2 Structural representations and interaction analyses of ligand binding within the protein. (**a–c**) 3D structures of the protein-ligand complexes for Ligand1, Ligand2, and Ligand3, respectively, showing ligand positioning within the binding pocket. (**d–f**) 2D interaction diagrams for Ligand1, Ligand2, and Ligand3 highlight key molecular interactions such as van der Waals forces, hydrogen bonds, π-π stacking, and alkyl interactions. Color coding illustrates the interaction types, emphasizing the binding stability and molecular contacts in each complex

hydrophobic interactions, hence augmenting the stability and binding affinity of the molecule. Collectively, these models provide a lucid depiction of the binding interactions, demonstrating that the ligand's orientation, position, and interactions with particular protein residues are essential to the binding affinity. This comprehensive study is highly beneficial for medicinal chemistry in the design of novel ligands with enhanced potency and selectivity.

10.3.4 Molecular Dynamics Simulations

Figure 10.3 presents an RMSD plot comparing the control (calpastatin) with three ligand-bound systems (Ligand1, Ligand2, and Ligand3) over time measured in picoseconds. RMSD is a crucial metric for assessing the structural variations and overall stability of biomolecular simulations concerning molecular complexes. For the control indicated by the black line, RMSD exhibits a gradual increase in the initial phases, then oscillating within the range of 0.35–0.45 Å after 10,000 picoseconds. This indicates that the control system undergoes minor structural modifications before returning to its equilibrium state while maintaining structural stability in the simulation [35]. Like the control, Ligand1 (Dihydroergotamine), depicted in orange, exhibits a rapid early increase followed by a decline to a plateau at

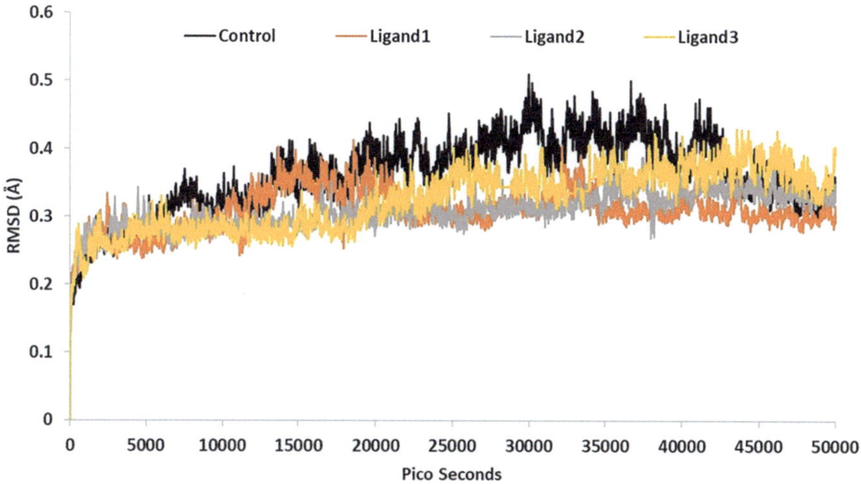

Fig. 10.3 RMSD profiles of the control and ligand-bound systems (Ligand1, Ligand2, and Ligand3) over 50,000 picoseconds. The graph shows structural deviations, indicating that Ligand2 exhibits the highest stability with the lowest RMSD, followed by Ligand1 and Ligand3, while the control displays more significant fluctuations

approximately 0.3–0.4 Å. This indicates that the binding of Ligand1 may restrict conformational alterations and enhance the stability of the complex. Ligand2 (Bolazine), depicted in gray, exhibits the lowest RMSD values (~0.25–0.35 Å) among all groups, signifying little structural variations over time. Consequently, the reduced RMSD suggests that Ligand2 may enhance the structural stability of the complex due to robust binding interactions with the site.

The yellow line illustrates that the RMSD for Ligand3 (Metergotamine) begins at approximately 0.3 to 0.45 Å, which is marginally more significant than that of Ligand1 and Ligand2, however comparable to the control level. This signifies moderate stabilizing effects, albeit with significantly greater flexibility than in Ligand2. The comparative RMSD profiles demonstrate distinct structural dynamics for each system, indicating that Ligand2 confers the most excellent stability to the protein, whilst Ligand1 and Ligand3 provide the subsequent highest stability. These discoveries may assist in guiding further research aimed at comprehending ligand binding mechanisms and their importance in molecular interactions.

Figure 10.4 illustrates the Root Mean Square Fluctuation (RMSF) for the control and three ligand-bound systems in relation to residue indices. RMSF provides insights into the flexibility of the backbone and the localized motion of individual residues; regions exhibiting more flexibility may correlate with loops, termini, or unstructured segments. The control group (black line) illustrates the variability value distributed over residues, with increased flexibility indicated by the peaks. The observed peaks likely result from loop regions or solvent-exposed residues, which often exhibit greater flexibility in molecular dynamics simulations.

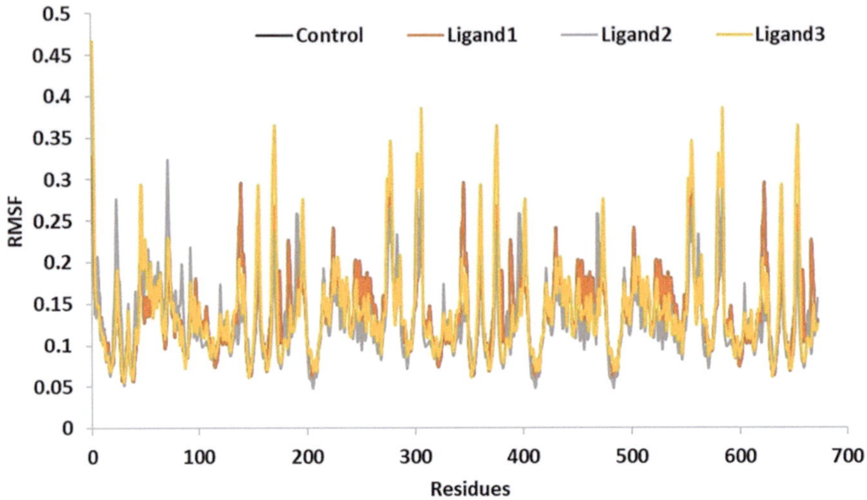

Fig. 10.4 RMSF profiles of the control and ligand-bound systems (Ligand1, Ligand2, and Ligand3) across residues. The data highlight regions of higher flexibility, with Ligand2 exhibiting slightly lower fluctuations, suggesting enhanced structural stability compared to the other systems

Ligand1 (orange) and Ligand3 (yellow) exhibit comparable oscillation patterns, with peak values and placements around those of the control. This observation indicates that these ligands do not influence the intrinsic plasticity of the protein structure and maintain dynamic properties akin to those of the unbound protein. Throughout the experiment, Ligand2 (gray) exhibited somewhat reduced oscillations in various places relative to the other systems. The reduction in flexibility may indicate increased rigidity, likely in specific residues participating in interactions with Ligand2, resulting in a more stable complex, as previously demonstrated.

The RMSF profiles demonstrate that all systems have a similar degree of dynamic flexibility; however, Ligand2 renders the protein structure somewhat more stable. These data suggest that Ligand2 may establish more substantial and more stable connections, perhaps enhancing structural stability while maintaining flexibility in the functional domains. Further investigation of the binding interactions may elucidate the molecular mechanisms behind these differences.

10.4 Discussion

The molecular modeling of the calpain receptor utilizing AlphaFold (AF-Q9HC96-F1-v4) has provided a robust basis for elucidating the topographical structure and functional dynamics of the receptor [5]. The model is appropriate for molecular docking and simulation investigations because of its high-confidence predictions of secondary structural features, binding pockets, and flexible loops, together with pLDDT scores. AlphaFold's structural data enables this model to deliver precise evaluations of receptor-ligand interactions, hence enhancing drug discovery efforts. Pocket predictions utilizing DeepBindPoc identified several critical residues, including Arg202, as the primary residues interacting with the ligand [11]. Utilizing GCNs enabled the identification of ligand-binding sites based on structural and chemical characteristics. The projected pockets coincided with conserved functional domains, so affirming their biological relevance. The data facilitated the creation of docking grids, and virtual screening of the receptor demonstrated its suitability for ligand binding, thereby confirming its utility in structure-based drug discovery [36].

Consequently, we utilized DeepBindGCN and AutoDock Vina for virtual screening to augment our understanding of ligand-receptor interactions. DeepBindGCN_BC and DeepBindGCN_RG produced excellent estimates of binding affinity, corroborated by negative binding energy values from AutoDock Vina. Samatasvir, Orvepitant, and Mk3207 have been discovered to possess the highest binding affinities and the most advantageous binding energy values, making them promising therapeutic candidates. Notably, Samatasvir exhibited a DeepBindGCN_RG of 9.52 and a binding energy of −10.60 kcal/mol, whereas Orvepitant demonstrated a marginally superior binding energy of −10.90 kcal/mol, suggesting robust binding affinities. The findings indicate that Abamectin-component-b1a serves as a binder across various entry and enhances its application in medication development. The molecular dynamics simulations provided further insight into the structural stability and dynamics of the ligand-receptor complexes. The analysis of RMSD values, considering protein structure and flexibility, indicated that Ligand2 exhibited the most advantageous RMSD values (~0.25–0.35 Å), signifying the compound's superior structural stability. Nevertheless, Ligand1 and Ligand3 exhibited marginally elevated RMSD values of approximately 0.3–0.4 Å, signifying modest stability compared to the control. It is suggested that Ligand2 induces minimal conformational alteration due to its robust binding and superior compatibility within the binding region.

A comprehensive evaluation of residue-level dynamics by RMSF analysis corroborated the conclusion about the differential impact of ligand binding on protein flexibility. Ligand2 exhibited a reduction in fluctuations across multiple areas of the protein and enhanced stiffness in the interacting residues. Ligand1 and Ligand3 exhibited flexibility profiles similar to the control, suggesting they do not impair intrinsic flexibility while providing minor stability. Collectively, these findings indicate the potential for enhancing structural stability through the use of Ligand2 while preserving functional versatility—an amalgamation that positions Ligand2 for further advancement. A synthesis of deep learning models and molecular dynamics simulations has been employed to assess the viability of therapeutic candidates and characterize their binding properties. The discovered compounds exhibited favorable docking scores and dynamic stability with the target protein, indicating their potential for therapeutic application. Future research will investigate these interactions in further depth through experimentation, improvement of lead compound structures, and the identification of additional ligands to enhance the chemical library. This study emphasizes the efficacy of computer-aided drug design and virtual screening techniques in contemporary drug development processes.

10.5 Conclusion

This study employed computational techniques, such as AlphaFold, DeepBindPoc, DeepBindGCN, and molecular dynamics simulations, to examine the structural characteristics and ligand-binding potential of the calpain receptor. The AI-predicted structural model demonstrated high confidence and accuracy for subsequent analysis. Pocket docking and virtual screening facilitated the identification of several potent ligand compounds exhibiting favorable binding affinities and stability. MD simulations validated that the configurations of ligand-receptor complexes were stable, with Ligand2 exhibiting the highest stability. Based on our discoveries, we firmly endorse the application of AI and deep learning algorithms in structural biology and drug discovery. These computational methods enable high-throughput screening and precise modeling, significantly reducing the time and cost associated with experimentation. Subsequent research will focus on the experimental validation of the specified ligands and the optimization of lead compounds for therapeutic agent development. This paper establishes a framework for employing computational methods to accelerate drug discovery and promote the investigation of protein-ligand interactions.

10.6 Data Availability

The data generated from the current study is presented in the paper.

References

1. Kairys V, et al. Recent advances in computational and experimental protein-ligand affinity determination techniques. Expert Opin Drug Discov. 2024;19(6):649–70.
2. Wang H. Prediction of protein–ligand binding affinity via deep learning models. Brief Bioinform. 2024;25(2):bbae081.
3. Wang DD, Chan M-T, Yan H. Structure-based protein–ligand interaction fingerprints for binding affinity prediction. Comput Struct Biotechnol J. 2021;19:6291–300.
4. Senthil R, et al. Bibliometric analysis of artificial intelligence in healthcare research: trends and future directions. Future Healthcare J. 2024;11(3):100182.
5. Borkakoti N, Thornton JM. AlphaFold2 protein structure prediction: implications for drug discovery. Curr Opin Struct Biol. 2023;78:102526.
6. Saravanan KM. The transformative potential of deep learning and alphafold in addressing brain disorders, vol. 3; 2024. p. 72–3.

7. Sazontova TG, Matskevich AA, Arkhipenko YV. Calpains: physiological and pathophysiological significance. Pathophysiology. 1999;6(2):91–102.
8. Croall DE, Ersfeld K. The calpains: modular designs and functional diversity. Genome Biol. 2007;8:1–11.
9. Metwally E, et al. Calpain signaling: from biology to therapeutic opportunities in neurodegenerative disorders. Front Vet Sci. 2023;10:1235163.
10. Zhang X, et al. Calpain: the regulatory point of cardiovascular and cerebrovascular diseases. Biomed Pharmacother. 2024;179:117272.
11. Zhang H, et al. DeepBindPoc: a deep learning method to rank ligand binding pockets using molecular vector representation. PeerJ. 2020;8:e8864.
12. Zhang H, Saravanan KM, Zhang JZH. DeepBindGCN: integrating molecular vector representation with graph convolutional neural networks for protein–ligand interaction prediction. Molecules. 2023;28(12):4691.
13. Feng Y, et al. Hybrid drug-screening strategy identifies potential SARS-CoV-2 cell-entry inhibitors targeting human transmembrane serine protease. Struct Chem. 2022;33(5):1503–15.
14. Skolnick J, et al. 2: why it works and its implications for understanding the relationships of protein sequence, structure, and function. J Chem Info Model. 2021;61:4827–31.
15. UniProt Consortium. The universal protein resource (UniProt). Nucleic Acids Res, Suppl_1. 2007;36:D190–5.
16. Jumper J, et al. Highly accurate protein structure prediction with AlphaFold. Nature. 2021;596(7873):583–9.
17. Berman HM, et al. The protein data bank. Nucleic Acids Res. 2000;28(1):235–42.
18. Lill MA, Danielson ML. Computer-aided drug design platform using PyMOL. J Comput Aided Mol Des. 2011;25:13–9.
19. Pettersen EF, Goddard TD, Huang CC, Couch GS, Greenblatt DM, Meng EC, Ferrin TE. UCSF chimera—a visualization system for exploratory research and analysis. J Comput Chem. 2004 Oct;25(13):1605–12.
20. Irwin JJ, et al. ZINC: a free tool to discover chemistry for biology. J Chem Inf Model. 2012;52(7):1757–68.
21. Lipinski CA. Lead-and drug-like compounds: the rule-of-five revolution. Drug Discov Today Technol. 2004;1(4):337–41.
22. Bento AP, et al. An open source chemical structure curation pipeline using RDKit. J Cheminformatics. 2020;12:1–16.
23. Zhang H, Saravanan KM. Advances in deep learning assisted drug discovery methods: a self-review. Curr Bioinforma. 2024;19(10):891–907.
24. Oleg OT, Olson AJ. AutoDock Vina: improving the speed and accuracy of docking with a new scoring function, efficient optimization, and multithreading. J Comput Chem. 2010;31(2):455–61.
25. Goodsell DS, et al. The AutoDock suite at 30. Protein Sci. 2021;30(1):31–43.
26. Baroroh U, et al. Molecular interaction analysis and visualization of protein-ligand docking using biovia discovery studio visualizer. Indonesian J Comput Biol (IJCB). 2023;2(1):22–30.
27. Pronk S, et al. GROMACS 4.5: a high-throughput and highly parallel open source molecular simulation toolkit. Bioinformatics. 2013;29(7):845–54.
28. Mark P, Nilsson L. Structure and dynamics of the TIP3P, SPC, and SPC/E water models at 298 K. Chem A Eur J. 2001;105(43):9954–60.
29. Jorgensen WL, et al. Comparison of simple potential functions for simulating liquid water. J Chem Phys. 1983;79(2):926–35.
30. Humphrey W, Dalke A, Schulten K. VMD: visual molecular dynamics. J Mol Graph. 1996;14(1):33–8.
31. Zhang H, et al. DeepBindRG: a deep learning based method for estimating effective protein–ligand affinity. PeerJ. 2019;7:e7362.
32. Zhang H, et al. DeepBindBC: a practical deep learning method for identifying native-like protein-ligand complexes in virtual screening. Methods. 2022;205:247–62.

33. Zhang H, et al. Artificial intelligence and computer-aided drug discovery: methods development and application. Methods (San Diego, Calif.). 2024:S1046–2023.
34. Sreeraman S, et al. Drug design and disease diagnosis: the potential of deep learning models in biology. Curr Bioinforma. 2023;18(3):208–20.
35. Zhang H, et al. An integrated deep learning and molecular dynamics simulation-based screening pipeline identifies inhibitors of a new cancer drug target TIPE2. Front Pharmacol. 2021;12:772296.
36. Zhang H, et al. Deep learning based drug screening for novel coronavirus 2019-nCov. Interdiscip Sci: Comput Life Sci. 2020;12:368–76.

Chapter 11
Bridging the Gap: Translational Opportunities of Graph Neural Networks from Research to Clinical Practice

Monirul Islam

11.1 Introduction

Interpretation is the transformation of findings in the laboratory, clinic, community, and into actions to enhance the health of the person and the population—from testing, treatment, and surgery to prevention and lifestyle adjustments. Translational research is the branch of study that targets the science and operations of each step of the process. Translation is the process of moving from in vitro and experimental animal studies to the humans. This entails routing basic experimental outcomes through to preclinical experiments, clinical trials, and final clinical usage. Translational research tries to develop more usable and impactful outcomes useful in improving human existence. Translational research aims at moving (translating) basic sciences discoveries more rapidly through the clinical interface. Translational research strictly speaking only means to facilitate as well as support interprofessional cooperation between research based in laboratories and clinics. It also considers the extent of its scope along with the general public by taking the communities to ask them what they want in terms of health innovations. And also, it defines and allows the adoption of overall sound medical and health responsibilities. Translational research is divided into waves according to which stage of translation process (from basic research to social utilization and outcomes) it is. T Spectrum or Translational Spectrum on the other hand comprises of different steps of translational research (T0- T4). Translational research is an area that can be greatly impacted by laboratory automation. Contemporary TM subsumes a whole spectrum of the biomedical research which refers to omics, pharmacology and drug discovery, cellular and molecular diagnostics, and animal models. Quite evidently, it remains relevant to make sure that these different approaches are capable of being coordinated to rapidly and effectively deliver the advantages of translational

M. Islam (✉)
Department of Management, Sister Nivedita University, Kolkata, West Bengal, India

M. M. Hassan et al. (eds.), *Graph Neural Networks for Neurological Disorders*,
https://doi.org/10.1007/978-3-032-04315-3_11

research and therefore, is now an aspect of global health policy, research, and funding agendas. This concept has also brought out several advances in the finding and exploration of diseases like the following through a translational research approach.

11.1.1 Infectious Diseases

COVID-19 exposed the critical role of an efficient bench-to-bedside process. From determining the genome of the SARS-CoV virus to the emergence of vaccines, the COVID-19 response articulated what can be done when everything starting from benchside to bedside processes functions as a unit.

11.1.2 Cancer

Recent advancements and innovations in molecular techniques that include omics technologies and biomarker prediction have increased the depth of converting findings observed in a laboratory to those that are applicable in cancer care. An example of this is the acknowledgement of biomarkers that helps determine the treatments required by Lung cancer patients—SLE or adenocarcinoma.

11.1.3 Genetic Diseases

Among the central tenets of translational research there was the Human Genome Project, a seminal effort that laid the groundwork for the utilization of genomic information for improved health. The objective of the project was to sequence out of all the human genomes and pin down genes that are involved in disease so that translational medicine therapies can be formulated.

The next segments of the chapter are organized into these sections. Section 11.2 provides a comprehensive overview of GNNs and their potential. Section 11.3 delves into the methodology employed in this study. Section 11.4 presents the foundations of GNNs which includes its core principles and architectures. Section 11.5 entails the GNN applications in Healthcare Research. Moreover, Sect. 11.6 discusses GNN applications in radiology. Section 11.7 deals with encoding patient EHRs. Section 11.8 talks about GNN importance in medical image denoising. Section 11.9 comprises the challenges in translational practice. Section 11.10 briefs potential future directions. Finally, in Section 11.11, the authors wrap up with the conclusion of the chapter.

11.2 Overview of GNNs and Their Potential

The neuron is a signal-performing neuro-organelle in the nervous system of an organism. Afterward, the information is received and then transmits signal through brain's synapses from one neuron to another until physiological effect is enacted [1]. ANNs are based on biological neural networks as computational methods that are built using algorithms for learning from experience [2]. Biological Neural Networks like Mammalian include billions of neurons and can be compared to ANNs having hundreds or thousands of units of process [3]. Also, as in biological networks, the learning abilities and the ability to iteratively optimize tasks performance is most valuable to ANNs. It is when knowledge is produced to such an extent that a solution is the generalized answer of some class of problems [4]. ANNs are very general-purpose and can be used for Engineering, Medical, General Science, and Economy [5]. In generality, GNN includes processing graph data by the deep learning algorithm. It can be applied either to the systems and communications of any organism which just innately has a graphical representation, i.e., social networking, physical models, financial bootstrap, and molecular modeling [6]. Nonetheless, through the use of graphs, GNNs aid with precision and faithful extrapolations as they do node feature operationalization as well as surrounding nodes and on the nature of their inter-relation with target [7]. The versatility is what has made GNN to solidify its position in this landscape over its initial nodes as graph kernels-based approaches [7]. It is not just for Sciences, the GNN has been developed to deal with Operation Systems or Management modeling and Analysis across multi-disciplinary research fields [8]. Due to its flexibility, GNN was able to be applied not only to these fields but also in Biology for being a successful virtual screening application GNN has been used for drug discovery in the field of area, also for discovering new proteins that might act as drug target sites [9]. Moreover, GNN imparts a key role in medical studies it aids refining the diagnostic and genetic surveillance of special diseases being used widely for handling products with a health focus, especially at the human and environment level [10].

A graph is a data structure in which we have a node. Few nodes are stated as vertices and edges. In math, it is relatively easy to cover as $G = (V, E)$ where nodes are not isomorphic E direction. These graph connections that we call edges can be either directed, representing a direction or value (in computational practice) that it traverses, or undirected, representing a number of. Graphs can be of different family data arrangements, for example, social data structural form, knowledge data structures, and protein data models. Graph space is one kind of non-Euclidean space (which is why the edge between two nodes in a graph can be incredibly distant from the distance between respective coordinates when we plot these on a Euclid space). The difficulty comes in to traditional neural networks when you feed graph data into them, they are the best at Euclidean optimization [11]. GNNs are a family of deep learning architectures implied to learn, extract models from graph data. GNNs do information aggregation from neighbors using message passing trick so that the model can learn structural associations out of phase. So far, we have seen GNNs useful in relation to node classification, link prediction and clustering, etc., for numerous tasks we observed GNNs research from 2005 till now. GNNs could model a more general graph and be used for node-specific applications, edge-specific applications, graph-specific applications, and so on. The concept of GNNs was introduced by Marco Gori in 2005 along with recursive neural network extended by GNNs [12]. In particular, GNNs are described as a subclass of neural nets, which have been developed to handle data in graph form. To establish interconnectivity and relationships, we use relational logic, where edges connect related nodes. GNNs are trained to learn both the node features and the underlying graph structure.

GNNs are the handmaiden of deep learning and an increasingly active area of research in AI and ML. The rapid development of GNNs has been made possible by the ubiquity within data in different contexts, like social media networks and biology, reference systems, and cybersecurity. GNN links are remarkable in modeling and comprehending intricate relational advances, which make them essential in solving actual-life complications which original machine-learning models cannot handle [13]. It is quite valuable that GNNs can model more structural information inherent in graph-structured data than other types of networks. Informing often appears as dependencies, connections, and contextual relations that are cogent for accurate prediction and decisive information. As a result, both GNNs and their applications have been developed and expanded across a range of problems, revolutionizing what is achievable in machine learning. In recent years, DL has been labeled as the most sophisticated technique of ML [14]. It has also gradually developed into the most common numerical method in ML and provides very good accuracy on numerous complex cognitive tasks occasionally even at the level of human beings. A major advantage found in DL is its ability to learn massive quantities of data [15]. Several modifications of GNN like GCNs, GATs, or GraphSAGE have been reported to yield remarkable results in different deep learning-based tasks in the recent past [8].

11.3 Methodology

This chapter followed a mixed-methods research design, alternating qualitative analyses on the functions and deployment of Graph Neural Networks (GNNs) in clinical practice. Central pillars to it are the systematic review of relevant literature, which considers the existing studies in GNN and also prior technological interventions that include discussion of the latest GNN and how we are using it to solve the gap between research ideas and real-world aspects in healthcare. The Case Study Analysis was applied to analyze real-life examples of GNNs in order to design new drug targets by modeling of disease evolution.

11.4 Foundations of GNNs

11.4.1 Core Principles and Architectures

- ***Message Communication Mechanism in Graph Neural Network***

Graph equilibriums are encoded in a GNN (node-edge-global context) permutation invariances of all graph properties as an optimizable transformation (permutation). To characterize the output is essentially the input in terms of similar adjacent list and the number of appeared vector count because GNNs preserve input connectivity [16]. As the GNN updates every node edge and global-context representation, the output has different embedding in graph but likewise the output graph also consists similar adjacent list and appeared vector count as input nodes. In the figure, rounds of it are circles are nodes and boxes in empty depict the aggregated neighbor or adjacency nodes. The model combinations message across A's typical local neighbors (in particular B, C, and D). Instead, the neighborhood from neighbors are generated messages by aggregation of their neighborhood alike. This visualization model shown in Fig. 11.1, which is greatly simplified, represents a two-layer message-passing model at the unrolled neighborhood around target node t, where the calculation graph of GNN constitutes a structure of tree [17]. GNNs refer to neural-based models which reflect the dependency of graphs through message communicating between nodes in a graph [12].

In a neural network, each node of the message-passing mechanism stores messages as appeared vectors and on each time the values are updated in feature vector form [18]. So, the grey color node is coupled to the blue color node by aggregating the information shown in Fig. 11.2. Together, both appeared features are coupled and generate appeared vectors after setting updated values with the new message.

GNN family is a family of neural network models developed particularly for working on graph-structured data-related information. They have been greatly lauded in many different areas, mostly because of their state-of-the-art performance on inferring complex relationships and structures in graph data as depicted in Fig. 11.3.

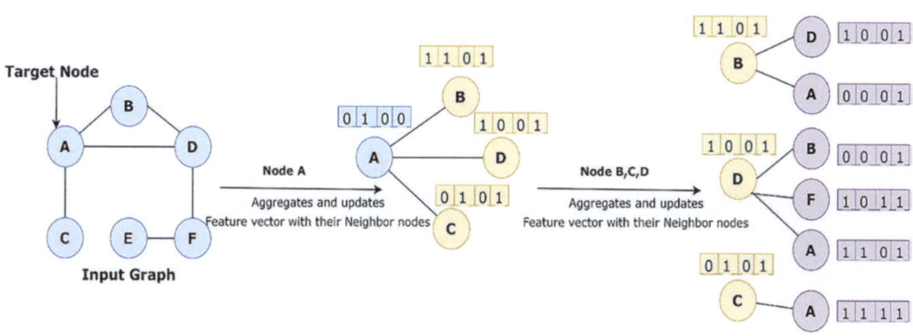

Fig. 11.1 How does a single node gather messages from its neighboring nodes

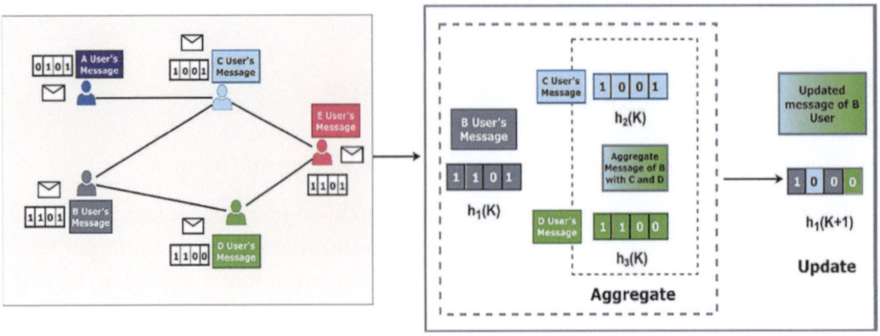

Fig. 11.2 GNN message-passing mechanism

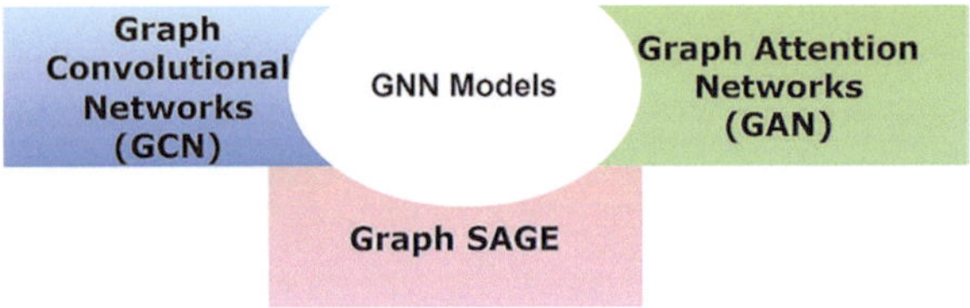

Fig. 11.3 Models of GNN (GCN, GAN, Graph SAGE)

- ***Graph Convolution Neural Network (GCN)***

GCN is a variant of simple GNN. GCN networks are a kind of GNN implemented by Thomas Kipf & Max Welling. Convolution layers are layers between Convolutional Neural Networks. Where neurons are multiplied by factors weights like filters or kernels. Filters are taken as gliding window on the image, to make CNN learn features from local cells. GCNs were first proposed in "Deep Locally Connected Networks on Graphs and Spectral Networks" [19]. GCNs behave similarly: they learn features by looking at neighboring nodes. The biggest difference between GNNs and CNNs is that CNNs are designed for regular (e.g., Euclidean) ordered data. In contrast, GNNs are a generalization of CNNs by having multiple nodes per layer and nodes not necessarily connected in an ordered fashion (on non-Euclidean structured data [20]. GCNs are an older deep model that performs incredibly well on problems that map onto graph data social networks/citation networks, recommender systems, etc. These networks are variations off the fairly common CNNs being applied to grid-like data (most notably images) as depicted in Fig. 11.4. At the heart of GCNs is the modification of convolution to work on graph data [21]. That allows them to both learn and spread over the layer of nodes in a network graph by combining node-related features as well as neighbor features. GCNs are usually multi-layered, where each layer consists of a convolution and an aggregation step that refines the node representations in the graphs. GCNs are agglomerative and propagative in the sense they aggregate information across nodes of the graph.

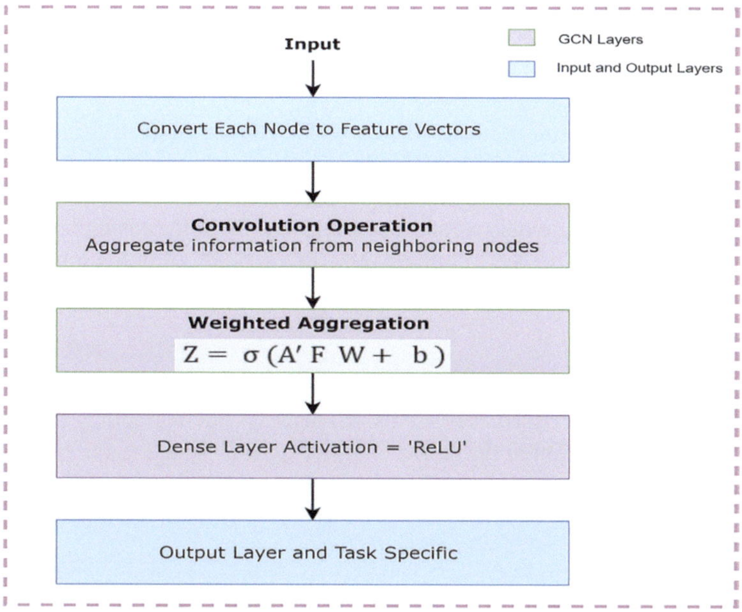

Fig. 11.4 Working model of GCN

- ***Graph Attention Network (GAT/GAN)***

GAT/GAN is a novel neural-based network technique that works on graph-based structured data. To remedy the limitations of prior methods that were either entirely based on graph convolutions or approximations thereof, it employs masked self-attentional layers. Through stacking layers, the process permits to be implicit (i.e., not having to assign weights to different nodes around individual unit distinctly) but yet enables nodes classify differently based on evidences of neighborhoods, forgoing to be nodes from performing expensive matrix operations like inversion without previous knowledge on the Graph. GAT takes care of several serious challenges that plagued the classical spectral-based GNN and thus can be used in both inductive and graph-agnostic applications. The GAT is such kind of Neural Network which solves both points from the figure. It uses attention in a neural network model on graphs. GATs are a set of GCN methods activated with attention mechanisms. From the Graph Convolutional Networks (GCN) that we saw earlier, it is of high performance for node classification to combine local graph structure and features of nodes. On the other hand, however, the way GCN aggregates messages depends on structure and hence should be susceptible as well.

11.4.2 Advantages of GNNs for Graph-Structured Data in Healthcare

GNN use cases in several domains including NLP, program analysis, data mining and comparison, bioinformatics, etc. GNN utilization will not happen without the first pre-processing or transformation/representation of input data into graph-like formats. GNNs are extremely good at learning node/graph embedding that are a crucial building block for the following tasks in the realm of graph representation learning.

- ***Molecules***

Atoms (made up of electrons) are the building blocks of matter and molecules are 3D graph-like nature species of atoms. All particles are interacting, but we identify a covalent connection in the atoms in case they can be said to far apart enough to be stable. There are single/double bonds(atom-to-atom) having different types.. Therefore, a 3D arrangement is symbolized as a graph where atoms correspond to related nodes and covalent bonds emanate between such nodes [22].

- ***Social Network Graphs***

Research tools to discern the collective behavior of majority social phenomena involving individuals and groups and their organizations. We can describe groups of individuals using a graph with individual nodes and edges connection [23].

- ***Scientometrics***

It is a regular occurrence for scientists to cite each other in their papers. Citation networks are where each manuscript is a node in the (directed) graph of citation research networks, and each publication credits the others. We can further embed each document in the node, e.g., word-level embedding as an abstract [5].

- ***In Computer Visualization***

For tagging specific points in visual scenes. So, then we would make graphs by considering those nodes and their edges as an interaction.

GNNs are used to model the data in graph form, where they naturally enable the representation of data with intricate relationships and dependences that conventional ML cannot easily model. This is why GNNs are so beneficial for the classes of problems in which the data has a natural structure, which is a graph, or when relationships must be modeled properly for accurate predictions/analysis.

11.5 Applications in Healthcare Research

It has applications in modeling patient networks and disease ontologies. This work addresses data-driven agricultural product management, immunization, and the study of vaccine adverse events. The fMRI configuration graph neural network led convolution in classification to discriminate MDD patients from controls.

11.5.1 Disease Ontologies

Identifying the background of diseases is essential in bioinformatics. To predict disease association prediction being handled the existing methods mainly include matrix decomposition methods, network propagation, and machine learning. In addition, some methods like machine learning are measures of similarity and matrix decomposition. However, the matrix factorization methods, which assign the features in entities to a latent space, are unable to capture the representation of the topological relationships between the entities. The shallow approaches that are utilized in several variations ignore the high-level structural information about entities in disease-related networks affecting the quality of entities' features. GNNs have recently taken the measure to deal directly with the non-linear relationship between disease-nodes from biological networks [24]. In most of these methods, the convolution operations were applied at local subgraphs level on heterogeneous network. Related research has fully or partially affected the progression of deep learning for disease prediction. These studies constructed various biochemical networks by integrating different types of information, such as RNA-disease associations, disease genes information, and so on.

11.5.2 RNA-Disease Association

A plethora of evidence demonstrated that microRNAs (miRNAs), long non-coding RNAs (lncRNAs), circRNAs, and Piwi-interacting RNA are one of the major regulatory parts in the pathogenesis and progression of diseases [25, 26]. Consequently, the identification of this RNA-disease association is essential for the pathogenesis of complex diseases. In a computational model-based framework, the RNA and disease data analysis can avoid the drawbacks of high biological experimental verification costs with substantial time-consuming tasks. Researchers started to introduce GNNs into this category of research in 2019. In the prediction of disease-associated RNAs with insufficient data, multi-subject information integration can be used to enhance our understanding of the complex interaction networks of biological computations. So, further investigation on how to learn the complex of near-interaction mode among multiple correlated data by integrating different data

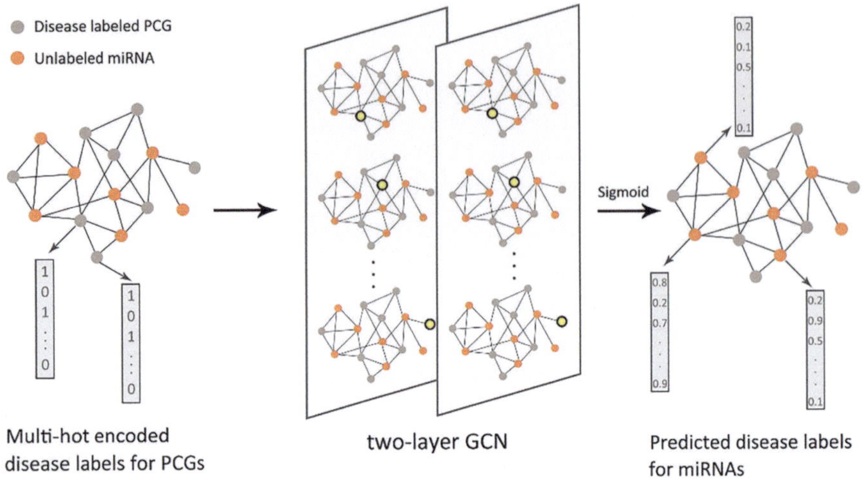

Fig. 11.5 Flow chart of DimiG in RNA-disease association

type is necessary for the graph deep learning model. This two-layer graph neural network took the protein-coding gene (PCG)-miRNA association network as the input. The PCGs were labeled, but the miRNAs were unlabeled. In the highlighted introduced ParOrange nodes, these yellow nodes were the weighted sum of neighbor embedding. The output could clue the correlation of probability between diseases and unlabeled miRNAs as shown in Fig. 11.5.

11.5.3 Disease-Gene Association

Single-cell RNA sequencing offers gene expression data from a single cell. GNNs are able to learn the cell-cell interaction and differentiate cell and disease state prediction [27]. Inference disease prediction from precise gene-disease association would be of great use for understanding the functions of disease-causing variants and offer evidence for possible prevention of diseases. It would be theoretically possible (modulo implementation and computational expense) to prioritize candidate genes for some diseases by addressing the DGP. Although considerable medical data has been salvaged in all sorts of databases, the pervasive prediction of cancer is a real hurdle. Cancer, as a class of diseases, is formed by different gene defects and exhibits synthetic lethal interaction among genes. Thus, the gene interaction network contributes significantly to cancer predictions. The interpretability of GNNs is also significantly improved through the correlation calculated for each data point and can be superior in molecular label predicting tasks that are related to patient-specific disease networks.

11.5.4 Diagnostics, Genomics, and Drug Discovery

- **Prediction of Protein-PP Interactions**

There are several GNN approaches that predict protein-protein interactions from the biological networks, which can help in drug discovery and understanding pathogenic mechanisms. DeepGraphGo has 3 features in Multispecies Protein Function Prediction: Representation vector InterPro, Multiple GCN layers, and Hoc strategy.

DeepGraphGO refers to the deep learning-based framework for unsupervised protein network reconstruction combining protein sequence and network data via Graph Neural Networks.

- **Genomic Sequence Analysis**

Gene interactions or genomic sequences are essential for gene expression prediction and sequence classification, which can be modeled by GNN. Molecular level prediction (infer from predict) LR-GNN combines in a GCN encoder that learns to get node embedded. An operator had been made to reflect relationships between molecules along propagation rule by LR representation encapsulated node embedded in each GCN-encoder layer and Link Representation (LR-GNN) as GNN applied model.

- **Drug Discovery**

Drug Development primarily encompasses drug-target identification, lead compound-related discovery, and optimization to candidate drug [28]. Because of the scarcity of drug targets, the low clinical translational value of animal models, and the diversity of patients/biological systems, development is a difficult job. Use case: Modern drug discovery approaches aim to reduce these middle steps by using machine learning for faster development and cost savings. Therefore, most of the researchers have a choice of preferring machine learning approaches rather than early-stage molecular interactions besides hand searching, which might greatly alleviate workload for the following experiments. One of the most concerning uses of machine learning methods, deep learning in biomedicine, is faced with a myriad of constraints. First of all: the vast majority of deep learning models are crafted to work on structured dataset complementing from the original input (require high-quality) and often has to be labeled; secondly, traditional GNNs and other deep models cannot be trained directly on such unstructured data (e.g., molecular graphs) such that the explicit representations of molecules are not well-defined by these models. Hence, GNNs—the method that generalizes deep learning to non-Euclidean domains—have been a burgeoning approach for drug-related tasks. GNNs are necessary to handle non-Euclidean drug-related tasks with respect to drug-to-target site interaction estimate and molecular-level property estimate, which are key in pharmaceutics study research.

The network is optimized for DPI prediction; that it goes through supervision signals from downstream task (Drug-Protein Interactions prediction) in the form of sampling. A GNN can provide insights at network level regarding many drugs and proteins in the drug-protein association network through information propagation within this network. It is a combination of network-level information with ML.

11.5.5 Integrating GNNs into Clinical Workflows

Proper explanations in AI and robust model validation are key to the successful adoption of GNNs. Researchers and clinicians must work together where these interdisciplinary teams can ensure a balance between technical advancements and clinical requirements. Common objectives, iterative feedback, and building user-friendly tools can unite AI and practice. In addition, training programs can guide the individuals who work with GNN-based tools and pilot studies can evaluate feasibility in real-world settings.

11.6 Applications in Radiology

The definition of disease diagnosis can be inferred as the process of determining what disease or condition is producing manifestations for the patient [29]. Medical imaging techniques like positron emission tomography (PET), diffusion-tensor imaging (DTI) and functional magnetic resonance imaging (fMRI) medical images have widely utilized for disease diagnosis. Medical image interpretation is a major occupation of radiologists and physicians, thus making diagnosis way too time-consuming. This problem is thus resolved with the advent of deep learning in disease diagnosis [30]. The success of CNNs relies on their ability to discover or learn representation structures in regular grid data through convolutional procedures. They can express the data structure regularity or pattern. At the same time, CNNs are just not built for structured and ordered data, and implementing complex features in irregular formations.

11.6.1 Autism Spectrum Disorder

ASD is a neurological disorder within the autism spectrum, characterized by communication difficulties, restricted interests, and repetitive behaviors [31]. Subjects in this database were symbolized as nodes and edges between nodes corresponded to the Pearson correlation of rsfMRI time-series. GCNs were then utilized to extract the significant patterns for similarity in graphs—multi-dimensional graphs from sMRI and fMRI. In each subject, they calculated the adjacencies matrix by using cosine similarity on sMRI. Then extracting feature vectors from fMRI to encode cranial information as a histogram. Lastly, their final fusion was expected to improve negatively using GCNs on the desired modalities of sMRI.

11.6.2 *Alzheimer's Disease*

Progressive, idiopathic, and unstable AD is a disease that ultimately weakens memory, cognitive faculties, and normal human behavior features by neurodegeneration. Dementia from AD in those aged 60 years and older people most typically develops from and affect about 50–75% patients. MCI and SMC form transitional phase in to normal cognition prognosis from AD. Hernandez et al. constructed graphs with biological interactions [32]. The results indicated that GNNs always outperform other machine learning methods for AD diagnosis based on the experiments. Cai et al. adopted GNN algorithms to learn the discriminative features from multimodal (sMRI and DTI) brain networks constructed by spatial relations, cross-modal association, and feature correlation [33]. They introduced a graph transformer-based geometric learning framework for multimodal (brain network) brain data estimating brain age. An AD diagnosis depends upon the difference in the estimated age and actual age, and is a significant biomarker. The irregular graph structures in medical images such as MRI contribute to the diagnosis of diseases. GNNs were also suggested to relieve from the challenges of spatial pattern exposing and understanding in irregular structures. The GNN-based approaches are being extensively utilized for the graphs to derive structure information to diagnostic diseases. Further some GNN based algorithms like GTN, scene graphs, and knowledge graphs also look promising for disease diagnosis.

11.7 Encoding Patient EHRs

GNNs are graph-structured data-specialized deep-learning architectures. In comparison to Convolutional Neural Networks (CNNs), the GNNs are for processing the complex relational data among nodes in the network as nodes are the nodes "passing messages" for them to be able to understand the graph dependencies. You find them in widely used Tensor Flow models. For instance, GAMENet employs machine learning through GCNs to acquire a medicine knowledge graph with information on drug-drug interactions (DDI) as well as co-occurrence. In clinical health data in EHRs, deep origins are rich with ontological and semantic relationships that also help uncover things like disease & diagnosis co-occurrences, some of which enrich patient-specific literature. Driven by the requirement of a framework accounting for the clinical relational data, we strive for an individualized medical KG that is spiraled for every admission medication as well as clinical data.

Link prediction on biomedical networks:

- Using GCN to extract node-aware features from the sequence and structural data.
- A GCN-based encoder was employed to retrain the node feats that take into consideration the inter-node dependency in a network.

- Train with GNN via Graph reconstruction by including pre-training the node features.

11.8 Importance in Medical Image Denoising

Medical image denoising is one of the essential elements in healthcare and medical sciences. There are at least several reasons why medical images need to be denoised.

Faster diagnosis and treatment—Medical images of X-rays, CT, MRI, and ultrasound are useful tools for medical practitioners to diagnose diseases and to develop treatment plans. A lot of the important structural details in those images will get washed out by noise, precluding small anomalies. Medical imaging de-noising through the denoising technology allows healthcare professionals to see clear and more precise anatomical structures, lesions, and other pathologic features in medical images. This would result in more informed diagnosis, advanced treatment planning, and finally result in better patient outcomes.

Decrease radiation exposure—X-ray and CT image denoising reduces the radiation dose that must be administered to acquire clean medical images. It is imperative to minimize radiation, especially for patients who will be getting repeated imaging, like cancer patients going under fractionated radiotherapy. There are denoising algorithms that provide good-quality, low-dose images that are radiologist-like but also minimize the risk of radiation exposure. This is to protect patients from being unnecessarily subjected to injury while getting the medical imaging needed.

Participate in research and training—Clean images are a sine qua non for medical research since denoised medical images are necessary to study diseases, design novel therapies, or process effectiveness status between treatments. Clear images are necessary in educating medical students and for training of physicians for image understanding/surgical planning on the part of medical professionals and researchers' denoise images for education.

11.8.1 Telemedicine and Teleconsultation

For example, as telemedicine and remote consultation become more relevant (especially in hard-to-reach or rural areas), denoising technology can help to improve the quality of medical images transmitted. Telemedicine is built on the idea that faraway specialists have more relevant expertise for specific image interpretations. If we remove noise from images prior to telemedicine, the experts are accurately making diagnoses and provide treatment recommendations. This is most essential in an emergency case and also when the specialized medical services are absent.

11.9 Challenges in Translation

Translational research can drive the development of new diagnostic test, therapies, and interventions into evidence-based practice with a focus on human health by transferring basic science to the clinic and improving patient care.

Though the translation of bench to bedside research has been slower than anticipated, some important hurdles still need to be cleared to bridge that gap between laboratory discoveries and new clinical interventions.

11.9.1 Low Reproducibility

It has been widely noted that many published findings cannot be reproduced due to non-robustness of the results. Pharmaceutical science is too dynamic a subject to changing characteristics that generate unexpected results from same research. Implications for translational research to translate pre-clinical findings into human studies are borne by industry and academia.

11.9.2 Translational Models

In vitro and animal models are typically used in translation research, but there are concerns about the extent to which these models resemble human circumstances because clinical trials often do not follow up with identical conclusions to in vitro and animal investigations.

11.9.3 Carrying the "Right" Investigation

The way basic science goes is who asks the questions and then which hypotheses will be generated. Translational research hypotheses are driven by how they could be used in clinical practice and not merely by asking whether interventions provide benefit or harm.

11.9.4 Research Portfolios

The type of research processes happening at the organizational and systems level and for use in translational research. This includes evidence-based regulatory and ethical research processes, patient recruitment, and bioinformatics access.

11.9.5 Research vs. Clinical Care

A split between science and medicine has also been noted and hopefully enabled with more collaboration, mostly across the levels of research. In the available datasets, there are issues with unbalanced datasets, very limited labeled data, related domain-specific data, and unwarranted datasets, probably from challenging labels at hand. Both of these are addressed by methods in transfer learning and domain adaptation. Typically, medical image data is pretty limited, but even labeling medical images is very expensive and time-consuming work. However, the lack of large-scale and high-quality annotated data is a significant challenge, especially for deep-learning models, which often require ample of labeled data.

11.9.6 How Accurate Are the Current Systems/Models

It takes a fed of deep learning GCN, GAT, and GraphSAGE to be able to give these models more accuracy and efficiency to the systems we have today. Furthermore, large domain-specific datasets for training models can drive performance. Another major hurdle is the enhancement of traditional text classification, well thoroughly treated with high-performance text classification accuracies and performance provided by advanced deep learning approaches and GNN.

11.9.7 Types of Noise and their Distribution

There are many types of noise in medical images, such as random noise, artefacts, motion artefacts, etc., and the distribution may be extremely complex. The design of graph neural networks for different noise types is a hard task, as the several classes of noise should be considered. They are also inherently quite difficult in computational complexity; graph neural networks generally have high computational demand, especially on hard graph structures and huge medical image datasets. Practically, while developing these models for the application some of them will need to be trained and deployed in devices with restricted computational resources. These practical matters should be taken into account when future in-depth study is done on the application of graph neural networks for medical image denoising.

The prediction of disease requires improvements in 3 aspects: evaluation of similarity of new nodes, introducing node attribute-based inputs, and assorted information handling. Broad similarity methods were used as majority research adopted in disease prediction. GNNs are applied to obtain the inductive knowledge of heterogeneous network with disease ontology similarity, RNA functional and relatable similarity and multiples of association data. However, establishment of different yet similarity of networks could make the structure of the GNN to be complex to a certain extent and seek an effective similarity assessment pattern for new diseases or new RNAs are necessary. Also caused more awareness toward the incorporation of node attributes in modeling, for example, disease ontology features and RNA structural and functional features to alleviate excess reliance on association information.

The Drug Development includes essentially drug target identification lead compound identification and optimization from hits (secondary scaffolding) to the target drug [34]. Progress is limited because of the infrequent development of drug targets, the low clinical translational value of animal models, and the diversity of patient and biological systems. So as a consequence, majority of researchers can be up in the preference machine learning type approaches over just early stage molecular interactions and drastically reduce the subsequent experiments' manual work. But among the most unsettling, methods use machine learning methods. Deep learning in biomedicine is limited by a plethora of constraints. Deep learning models are highly customized with most deep learning models are designed to work on a structured dataset prepped from the original input (usually requires high quality and often has to be annotated), two; traditional GNNs and deepest models cannot be readily trained on raw unstructured data (e.g., graphs of molecules) as the explicit representations of molecules are not well-defined by these models. Therefore, the generalization of deep learning to non-Euclidean domains is being viewed as an emergent approach for drug-related tasks.

Deep learning black-box problem is interpreted as something that needs to survive. Since usually, entities and relationships of GNNs are in essence different types of real-word objects. GNNs have more capacity of being interpretable analyses, and will produce interpretable visualizations. Interpretation of GNNs for modeling biology networks constitutes a promising and challenging research direction.

11.10 Future Directions

Possible future directions for GNNs are needed to fill in the areas of current models that are shortcoming, which is key in order to advance their capabilities from accurate and reliable to even better by learning things like your ideal number of layers, or at least the graphs you will be operating on its heterogeneity & dynamics in resonance. Research on enhancing datasets, the understand ability of GNN model and attention mechanisms will lead us not only to insight into their decision boundary but also suggest future research directions. One example of this is the GNN and Explainable AI also abbreviated as XAI, which consists of methods and techniques to make decisions taken by an AI model understandable for humans giving clear explanations about how the specific reasons a model of AI made certain decision (You et al., 2021). The link between XAI and GNN is critical as GNNs have such a complex architecture to understand. It will also make available information on how these networks operate in deciding for XIA on GNNs (thus becoming crucial for verification purposes and avoiding analyzing misinterpreted results) and allow recycling the whole GNN after some modification, due to user demand [35].

The research gaps pointed out are lack of available ready datasets, inconsistent and varying datasets and disorganized and inefficient pipeline for each features extraction so far. This can be addressed with a targeted Domain Adaptation. In order to enhance Text Classification, a combination run deep learning and machine learning is required and the GNNs that are needed in this formulation increase classification accuracy. Discovery of Existing Systems (How accurate the existing detection systems are) and Type of the structure (Homogeneous as well as heterogeneous) can be improved by applying deep learning architectures, i.e., GCN, GAT, or GraphSAGE.

11.11 Conclusion

In a word, Graph Neural Networks (GNN) have been making remarkable progress in dealing with peculiar problems from structured data viz. a field where traditional and sloth deep learning methods, baked for images and text usually fail at providing any meaningful insights. GNNs provide the strong and general-purpose solution for applications built on top of graph structures. This broad research on GNNs goes into details explaining some of the most important points like GNN basics, relationship with ConvNets, message passing in GNN, use cases of GNN models, and possibilities in the future. In the core of our discussion are indeed explaining basic GNN, a hot topic with multiple application today that actually prop for and against each other, our understanding of this tech is converging. GNN, the subgroup of deep learning in non-Euclidean spaces, has a superior performance for handling graph-structured data on several tasks. As of now, GNN has produced very good results in (mostly) low-resource tasks. These present a number of challenges about data quality processing, methods, and interpretability, that could take a few long years to overcome and if the road is a little cloudy for now.

In this chapter, we summarize the translational opportunities provided by GNNs and their implications to healthcare. Through the powerful application of GNNs, diagnostic accuracy, treatment personalization, and general healthcare efficiency can all be transformed for researchers and clinicians alike. This combination is powerful as GNNs allow us to model complex biological systems and explore new avenues in drug discovery and patient-specific treatment approaches. In addition, handling multimodal data allows for a comprehensive view of health, promoting precision medicine and preventive measures. Key are collaborative efforts between computational experts and healthcare providers to overcome implementation barriers. As research progresses and multi-disciplinary team efforts continue to evolve, GNNs can be transformative in personalized healthcare reaching from bench to bedside. A Call for Action to Leverage Graph Neural Networks for Interdisciplinary Collaboration in Healthcare Researchers, care providers, data scientists, and policymakers need to collaborate to harmonize technological innovation with clinical realities. Collaborative partnerships between person in computation and medicine can help accelerate translating research breakthroughs into diagnostic or therapeutic tools that ultimately better clinical outcomes. From structured forums for the exchange of knowledge, inter-disciplinary training programs and joint pilot projects can help facilitate a deeper understanding of mutual goals and challenges. This chapter, therefore, helps to build an ecosystem of collaboration that can enrich the deployment of such solutions in real clinical use with an emphasis on impact and sustainability to further advance the quality and accessibility of modern medicine. We think that GNNs are possibly an awesome computational tool to solve many of the biological problems within bioinformatics research. Finally, this chapter will be a valuable resource for upcoming researchers who will be working on the things discussed in this chapter.

References

1. Atz K, Grisoni F, Schneider G. Geometric deep learning on molecular represen-tations. Nat Mach Intell. 2021;3:1023–32.
2. Carron-Arthur B, Reynolds J, Bennett K, Bennett A, Cunningham JA, Grif-fiths KM. Community structure of a mental health internet support group: modularity in user thread participation. JMIR Mental Health. 2016;3:e20.
3. Giraldo JH, Skianis K, Bouwmans T, Malliaros FD. On the trade-off between over-smoothing and over-squashing in deep graph neural networks. In: CIKM'23: Proceedings of the 32nd ACM International Conference on Information and Knowledge Management, Birmingham, UK, 21–25 October 2023. New York: ACM. p. 566–76.
4. Feldman AM. Bench-to-bedside; clinical and translational research; personal-ized medicine; precision medicine—what's in a name? Clin Trans Sci. 2015;8(3):171–3.
5. Hodges H, Fealko C, Soares N. Autism spectrum disorder: definition, epidemiolo-gy, causes, and clinical evaluation. Transl Pediatr. 2020;9(Suppl 1):55–65.
6. Metsch JM, Saranti A, Angerschmid A, Pfeifer B, Klemt V, Holzinger A, Hauschild A-C. CLARUS: an interactive explainable AI platform for manual counterfactuals in graph neural networks. J Biomed Inform. 2024;150:104600.
7. Jain AK, Mao J, Mohiuddin KM. Artificial neural networks: a tutorial. Computer. 1996;29:31–44.
8. Corso G, Stark H, Jegelka S, Jaakkola T, Barzilay R. Graph neural net-works. Nat Rev Methods Primers. 2024;4(1):17.
9. Díaz-Rodríguez N, Herrera F. Explainable artificial intelligence (XAI): what we know and what is left to attain trustworthy artificial intelligence. Inf Fusion. 2023;99:101850.
10. Hasson U, Nastase SA, Goldstein A. Direct fit to nature: an evolutionary per-spective on biological and artificial neural networks. Neuron. 2020;105:416–34.
11. Borgatti SP, Everett MG, Johnson JC. Analyzing social networks. 2nd ed. Thousand Oaks: SAGE Publications Ltd.; 2018.
12. Gori M, Monfardini G, Scarselli F. A new model for earning in graph domains. Proc Int Jt Conf Neural Networks. 2005;2(2005):729–34.
13. Fan W, et al. Graph neural networks for social recommendation. In: The world wide web conference; 2019. p. 417–26.
14. He Q, Li X, Cai B. Graph neural network recommendation algorithm based on improved dual tower model. Sci Rep. 2024;14(1):3853.
15. Blondel VD, Guillaume J-L, Lambiotte R, Lefebvre E. Fast unfolding of communities in large networks. J Stat Mech Theory Exp. 2008:P10008.
16. Nerurkar PA, Chandane M, Bhirud S. Exploring convolutional auto-encoders for representation learning on networks. Comput Sci. 2019;20:273–88.

17. Eschmann R, Groshek J, Li S, Toraif N, Thompson JG. Bigger than sports: identity politics, Colin Kaepernick, and concession making in #BoycottNike. Comput Human Behav. 2021;114:106583.
18. Fudge N, Sadler E, Fisher HR, et al. Optimising translational research Opportuni-ties: a systematic review and narrative synthesis of basic and clinician Scien-tists' perspectives of factors which enable or hinder translational research. PLoS One. 2016;11(8):e0160475.
19. Shen D, Wu G, Suk HI. Deep learning in medical image analysis. Annu Rev Biomed Eng. 2017;19(1):221–48.
20. Wu Z, Pan S, Chen F, Long G, Zhang C, Yu PS. A comprehensive survey on graph neural networks. IEEE Trans Neural Network Learn Sys. 2019;32:4–24.
21. Schneider L, Guo Y, Birch D, Sarkies P. Network-based visualisation reveals new insights into transposable element diversity. Mol Sys Biol. 2021;17:e9600.
22. Pucci A, Gori M, Hagenbuchner M, Scarselli F, Tsoi AC. Investigation into the ap-plication of graph neural networks to large-scale recommender systems. Infona.pl. 2006;32(4):17–26.
23. Mahmud FB, Rayhan MM, Shuvo MH, Sadia I, Morol MK. A comparative anal-ysis of graph neural networks and commonly used machine learning algorithms on fake news detection. Proc – 2022 7th Int Conf Data Sci Mach Learn Appl CDMA. 2022:97–102.
24. Xia L, Chen B, Hunt K, Zhuang J, Song C. Food safety awareness and opin-ions in China: a social network analysis approach. Foods. 2022;11:2909.
25. Li Y, Qian B, Zhang X, Liu H. Graph neural network-based diagnosis predic-tion. Big Data. 2020;8:379–90.
26. Yan W, Zhang Z, Zhang Q, Zhang G, Hua Q, Li Q. Deep data analysis-based agricultural products management for smart public healthcare. Front Public Health. 2022;10:847252.
27. Krogh A. What are artificial neural networks? Nat Biotechnol. 2008;26:195–7.
28. Waikhom L, Patgiri R. Graph neural networks: methods, applications, and opportunities. arXiv. 2021;
29. Zhang Z, Chen L, Zhong F, Wang D, Jiang J, Zhang S, Jiang H, Zheng M, Li X. Graph neural network approaches for drug-target interactions. Curr Opin Struct Biol. 2022;73:102327.
30. Richens JG, Lee CM, Johri S. Improving the accuracy of medical diagnosis with causal machine learning. Nat Commun. 2020;11(1):3923.
31. Choi KS, Kim S, Kim B-H, Jeon HJ, Kim J-H, Jang JH, Jeong B. Deep graph neural network-based prediction of acute suicidal ideation in young adults. Sci Rep. 2021;11:15828.
32. Hernández-Lorenzo L, Hoffmann M, Scheibling E, List M, Matías-Guiu JA, Aya-la JL. On the limits of graph neural networks for the early diagnosis of Alzheimer's disease. Sci Rep. 2022;12(1):17632.
33. Cai H, Gao Y, Liu M. Graph transformer geometric learning of brain networks us-ing multimodal MR images for brain age estimation. IEEE Trans Med Imag. 2022;42(2):456–66.
34. Zhou J, et al. Graph neural networks: a review of methods and applications. AI Open. 2020;1:57–81.
35. Wu X-G, Wu H-J, Zhou X, Zhao X, Lu K. Towards defense against adver-sarial attacks on graph neural networks via calibrated co-training. J Comput Sci Technol. 2022;37:1161–75.

Index

M. M. Hassan et al. (eds.), *Graph Neural Networks for Neurological Disorders*, https://doi.org/10.1007/978-3-032-04315-3

MIX
Papier aus verantwortungsvollen Quellen
Paper from responsible sources
FSC® C105338

If you have any concerns about our products,
you can contact us on
ProductSafety@springernature.com

In case Publisher is established outside the EU,
the EU authorized representative is:
Springer Nature Customer Service Center GmbH
Europaplatz 3, 69115 Heidelberg, Germany

Printed by Libri Plureos GmbH
in Hamburg, Germany